REVIEW OF THE ARMY NON-STOCKPILE
CHEMICAL MATERIEL DISPOSAL PROGRAM

Disposal of Chemical Agent Identification Sets

COMMITTEE ON REVIEW AND EVALUATION OF THE ARMY
NON-STOCKPILE CHEMICAL MATERIEL DISPOSAL PROGRAM

BOARD ON ARMY SCIENCE AND TECHNOLOGY

COMMISSION ON ENGINEERING AND TECHNICAL SYSTEMS

NATIONAL RESEARCH COUNCIL

NATIONAL ACADEMY PRESS
Washington, D.C. 1999

NOTICE: The project that is the subject of this report was approved by the Governing Board of the National Research Council, whose members are drawn from the councils of the National Academy of Sciences, the National Academy of Engineering, and the Institute of Medicine. The members of the committee responsible for the report were chosen for their special competencies and with regard for appropriate balance.

The National Academy of Sciences is a private, nonprofit, self-perpetuating society of distinguished scholars engaged in scientific and engineering research, dedicated to the furtherance of science and technology and to their use for the general welfare. Upon the authority of the charter granted to it by the Congress in 1863, the Academy has a mandate that requires it to advise the federal government on scientific and technical matters. Dr. Bruce Alberts is president of the National Academy of Sciences.

The National Academy of Engineering was established in 1964, under the charter of the National Academy of Sciences, as a parallel organization of outstanding engineers. It is autonomous in its administration and in the selection of its members, sharing with the National Academy of Sciences the responsibility for advising the federal government. The National Academy of Engineering also sponsors engineering programs aimed at meeting national needs, encourages education and research, and recognizes the superior achievements of engineers. Dr. William A. Wulf is president of the National Academy of Engineering.

The Institute of Medicine was established in 1970 by the National Academy of Sciences to secure the services of eminent members of appropriate professions in the examination of policy matters pertaining to the health of the public. The Institute acts under the responsibility given to the National Academy of Sciences by its congressional charter to be an adviser to the federal government and, upon its own initiative, to identify issues of medical care, research, and education. Dr. Kenneth I. Shine is president of the Institute of Medicine.

The National Research Council was organized by the National Academy of Sciences in 1916 to associate the broad community of science and technology with the Academy's purposes of furthering knowledge and advising the federal government. Functioning in accordance with general policies determined by the Academy, the Council has become the principal operating agency of both the National Academy of Sciences and the National Academy of Engineering in providing services to the government, the public, and the scientific and engineering communities. The council is administered jointly by both Academies and the Institute of Medicine. Dr. Bruce M. Alberts and Dr. William A. Wulf are chairman and vice chairman, respectively, of the National Research Council.

This is a report of work supported by Contract DAAG55-98-C-0046 between the U.S. Army and the National Academy of Sciences. Any opinions, findings, conclusions, or recommendations expressed in this publication are those of the author(s) and do not necessarily reflect the view of the organizations or agencies that provided support for the project.

International Standard Book Number 0-309-06879-7

Limited copies are available from:

Board on Army Science and Technology
National Research Council
2101 Constitution Avenue, N.W.
Washington, DC 20418
(202) 334-3118

Additional copies are available for sale from:

National Academy Press
Box 285
2101 Constitution Ave., N.W.
Washington, DC 20055
(800) 624-6242 or (202) 334-3313
(in the Washington Metropolitan area)
http://www.nap.edu

Copyright 1999 by the National Academy of Sciences. All rights reserved.
Printed in the United States of America.

COMMITTEE ON REVIEW AND EVALUATION OF THE ARMY NON-STOCKPILE CHEMICAL MATERIEL DISPOSAL PROGRAM

JOHN B. CARBERRY, chair, E.I. DuPont de Nemours and Company, Wilmington, Delaware
JOHN C. ALLEN, ICF Kaiser, Boston, Massachusetts
LISA M. BENDIXEN, Arthur D. Little, Inc., Cambridge, Massachusetts
JUDITH A. BRADBURY, Pacific Northwest National Laboratory, Washington, D.C.
MARTIN C. EDELSON, Ames Laboratory, Ames, Iowa
SIDNEY J. GREEN, TerraTek, Inc., Salt Lake City, Utah
PAUL F. KAVANAUGH, consultant, Fairfax, Virginia
DOUGLAS M. MEDVILLE, MITRE (retired), Reston, Virginia
JAMES W. MERCER, HSI GeoTrans, Inc., Sterling, Virginia
WINIFRED G. PALMER, Henry M. Jackson Foundation for the Advancement of Military Medicine, Frederick, Maryland
GEORGE W. PARSHALL, E.I. DuPont de Nemours and Company (retired), Wilmington, Delaware
JAMES P. PASTORICK, GEOPHEX UXO, Alexandria, Virginia
WILLIAM J. WALSH, Pepper Hamilton LLP, Washington, D.C.
RONALD L. WOODFIN, Sandia National Laboratories, Albuquerque, New Mexico

Board on Army Science and Technology Liaison

E.R. (VALD) HEIBERG, III, Heiberg Associates, Inc., Mason Neck, Virginia

Staff

TRACY D. WILSON, study director (until 6/99)
MICHAEL A. CLARKE, study director
GREG EYRING, consultant
ROBERT J. KATT, consulting technical writer
HARRISON T. PANNELLA, research associate
SHIREL R. SMITH, senior project assistant (until 3/99)
DELPHINE D. GLAZE, administrative assistant (since 3/99)
MARGO L. FRANCESCO, publication manager

BOARD ON ARMY SCIENCE AND TECHNOLOGY

WILLIAM "BUD" H. FORSTER, chair, Northrop Grumman Corporation, Baltimore, Maryland
THOMAS L. MCNAUGHER, vice chair, RAND Corporation, Washington, D.C.
RICHARD A. CONWAY, Union Carbide Corporation, Charleston, West Virginia
GILBERT S. DECKER, consultant, Los Gatos, California
ROBERT J. HEASTON, Guidance and Control Information Analysis Center (retired), Naperville, Illinois
E.R. (VALD) HEIBERG III, Heiberg Associates, Inc., Mason Neck, Virginia
GERALD J. IAFRATE, University of Notre Dame, Notre Dame, Indiana
KATHRYN V. LOGAN, Army Research Office, Research Triangle Park, North Carolina
JOHN H. MOXLEY III, Korn/Ferry International, Los Angeles, California
STEWART D. PERSONICK, Drexel University, Philadelphia, Pennsylvania
MILLARD F. ROSE, NASA Marshall Space Flight Center, Huntsville, Alabama
GEORGE T. SINGLEY III, Hicks & Associates, Inc., McLean, Virginia
CLARENCE G. THORNTON, Army Research Laboratories (retired), Colts Neck, New Jersey
JOHN D. VENABLES, Venables and Associates, Towson, Maryland
JOSEPH J. VERVIER, ENSCO, Inc., Melbourne, Florida
ALLEN C. WARD, Ward Synthesis, Inc., Ann Arbor, Michigan

Staff

BRUCE A. BRAUN, director
MICHAEL A. CLARKE, associate director
MARGO L. FRANCESCO, administrative associate
DEANNA SPARGER, senior project assistant
ALVERA V. WILSON, financial associate

Preface

The Committee on Review and Evaluation of the Army Non-Stockpile Chemical Materiel Disposal Program (see Appendix A) was appointed by the National Research Council (NRC) to conduct studies on technical aspects of the U.S. Army Non-Stockpile Chemical Materiel Disposal Program. During its first year, the committee has evaluated a U.S. Department of Defense assessment of plans for the disposal of chemical agent identification sets—test kits used for soldier training. In its second year, the committee will provide recommendations on the midterm plans for the non-stockpile disposal program.

During its initial meetings, the committee received a number of briefings (see Appendix B) and held subsequent deliberations. The committee is grateful to the many individuals, particularly Colonel Edmund W. ("Ned") Libby, project manager for Non-Stockpile Chemical Materiel, and his staff, who provided technical information and insights during these briefings. This information provided a sound foundation for the committee's work.

This study was conducted under the auspices of the NRC's Board on Army Science and Technology. The committee acknowledges the support of Director Bruce A. Braun, and the board staff. The chair is also particularly grateful to the members of this committee, who along with the study director, the committee support staff, and the publication staff, worked diligently and effectively on a demanding schedule to produce this report.

John B. Carberry, chair
Committee on Review and Evaluation of the
Non-Stockpile Chemical Materiel Disposal Program

Acknowledgments

This report has been reviewed by individuals chosen for their diverse perspectives and technical expertise, in accordance with procedures approved by the NRC's Report Review Committee. The purpose of this independent review is to provide candid and critical comments that will assist the authors and the NRC in making the published report as sound as possible and to ensure that the report meets institutional standards for objectivity, evidence, and responsiveness to the study charge. The content of the review comments and the draft manuscript remain confidential to protect the integrity of the deliberative process. We wish to thank the following individuals for their participation in the review of this report:

William B. Bacon, consultant
John C. Bailar, University of Chicago
Joan B. Berkowitz, Farkas Berkowitz & Company
P.L. Thibaut Brian, (retired) Air Products and Chemicals, Inc.
Peter L. deFur, Virginia Commonwealth University
Elisabeth M. Drake, Massachusetts Institute of Technology
Donald E. Gardner, Inhalation Toxicology Associates
David S. Kosson, Rutgers, The State University of New Jersey
David P. Rall, (retired) National Institute of Environmental Health Sciences
Michael J. Ryan, Bechtel Jacobs Co., LLC
Barry M. Trost, Stanford University

While the individuals listed above have provided many constructive comments and suggestions, responsibility for the final content of this report rests solely with the authoring committee and the NRC.

Contents

EXECUTIVE SUMMARY 1
    Classification and Regulation of CAIS for Transport and Disposal, 2
    Commercial Incineration, 3
    Rapid Response System, 6
    Nonincineration-Based Options, 7
    A Path Forward, 8

1 INTRODUCTION 11
    Statement of Task and Congressional Direction, 12
    Chemical Agents and CAIS, 14
    Programs for Disposing of CAIS and Other Chemical Warfare Materiel, 22
    Legal and Regulatory Context for CAIS Disposal, 28
    International Approaches to CAIS Disposal, 29
    CAIS and the Chemical Weapons Convention, 30

2 DISPOSAL ALTERNATIVES 31
    Alternatives Considered, 31
    Alternatives Selected for Analysis, 37

3 ISSUES TO CONSIDER 38
    Technology, 38
    Laws and Regulations, 39
    Costs, 41
    Environmental Impacts, Worker/Public Safety, and Risks, 43
    Public/Stakeholder Involvement, 47
    Programmatic Considerations, 51

4 REVIEW OF THE COMMERCIAL INCINERATION OPTION 53
    Technology, 53
    Laws and Regulations, 59
    Costs, 60
    Environmental Impacts, Worker/Public Safety, and Risks, 65
    Public/Stakeholder Involvement, 67
    Programmatic Considerations, 73

5 ALTERNATIVES TO COMMERCIAL INCINERATION OF CAIS 75
    Baseline, Mobile Rapid Response System, 75
    Fixed Rapid Response System, 82
    Nonincineration Alternatives, 86

6 CONCLUSIONS AND RECOMMENDATIONS 94
    Classification and Regulation of CAIS for Transport and Disposal, 94
    Commercial Incineration, 98
    Rapid Response System, 101
    Nonincineration-Based Options, 103
    A Path Forward, 103

REFERENCES 105

APPENDICES

    A   Biographical Sketches of Committee Members, 113
    B   Committee Meetings and Other Activities, 117
    C   Methods of Treating Non-Stockpile Chemical Materiel, 122
    D   Legal Context for CAIS Disposal, 125

Figures, Tables, and Boxes

FIGURES

1-1 Army photographs of four CAIS types, 19

2-1 CAIS disposal alternatives, 32

3-1 Comparison of acute lethal concentrations of CAIS chemicals and some highly toxic industrial chemicals, 46

TABLES

1-1 Chemical Names and Formulas of CAIS Chemicals, 15
1-2 Characteristics and Biological Effects of CAIS Chemicals, 16
1-3 CAIS Types and Components, 20
1-4 Recovered CAIS Currently in Storage, 23
1-5 Potential CAIS Burial Sites, as Reported to Congress by the Army, 24
1-6 Status of Agent Destruction at JACADS and TOCDF, as of April 25, 1999, 25

2-1 Commercial Incinerator Facilities with Hazardous Waste Permits, 34

3-1 Properties of Sulfur Mustard and Lewisite, 44

4-1 Summary Evaluation of the Commercial Incineration Option, 54

5-1 Summary Evaluation of the Mobile RRS Option, 76
5-2 Summary Evaluation of the Fixed RRS Option, 84
5-3 Summary Evaluation of Selected Nonincineration Options, 87

6-1 Summary Evaluation of all CAIS Disposal Options, 95

BOXES

1-1   Use of CAIS, 18

3-1   Case Study: CAIS Recovery at the Raritan Arsenal, 40
3-2   Workplace Exposure Standards, 45
3-3   Risk Analysis Process, 48
3-4   Assessing the Public Acceptability of CAIS Disposal Options, 50

5-1   Expedient CAIS Disposal System, 83

Acronyms and Abbreviations

ACAMS	Automatic Chemical Agent Monitoring System
ACWA	Assembled Chemical Weapons Assessment Program
ASME	American Society of Mechanical Engineers
ATA	Alternative Technologies and Approaches Program
CAIS	chemical agent identification sets
CAMDS	Chemical Agent Munitions Disposal System
CERCLA	Comprehensive Environmental Response Compensation and Liability Act
CWC	Chemical Weapons Convention
CWM	chemical warfare materiel
DAAMS	Depot Area Air Monitoring System
DCD	Deseret Chemical Depot
DRE	destruction and removal efficiency
ECS	Expedient CAIS Disposal System
EDS	Emergency Destruction System
EPA	Environmental Protection Agency
GB	sarin (nerve agent)
HD	sulfur mustard, distilled
JACADS	Johnston Atoll Chemical Agent Disposal System
MEA	monoethanolamine
MMAS	Multiple Munitions Assessment System
MINICAMS	Miniature Continuous Air Monitoring System
MMD	munitions management device
NEPA	National Environmental Policy Act
NRC	National Research Council
NSCMP	Non-Stockpile Chemical Materiel Program

OSHA	Occupational Safety and Health Administration
PCB	polychlorinated biphenyl
PIG	package in-transit gas shipment
PINS	portable isotopic neutron spectroscopy
RCRA	Resource Conservation and Recovery Act
RRS	Rapid Response System
SCWO	supercritical water oxidation
TDG	thiodiglycol
TOCDF	Tooele Chemical Agent Disposal Facility
TSDF	treatment, storage, and disposal facility
USC	United States Code
UXO	unexploded ordnance
VX	a nerve agent

Executive Summary

This study is a review and evaluation of the U.S. Army's *Report to Congress on Alternative Approaches for the Treatment and Disposal of Chemical Agent Identification Sets (CAIS)*. CAIS are test kits that were used to train soldiers from 1928 to 1969 in defensive responses to a chemical attack. They contain samples of chemicals that had been or might have been used by opponents as chemical warfare agents. The Army's baseline approach for treating and disposing of CAIS has been to develop a mobile treatment system, called the Rapid Response System (RRS), which can be carried by several large over-the-road trailers.

In 1997, Congress directed the U.S. Department of Defense to assess the existing policy and plans for disposing of CAIS and report on disposal alternatives and policy changes that "could result in significant reductions in the cost . . . with no reduction in overall program safety." The Army's report, which responds to this congressional mandate, focuses on the alternative of shipping CAIS to existing commercial facilities for treating and disposing of hazardous wastes. This "commercial disposal option" has the potential to cost much less than deploying an RRS to each site where CAIS are stored or where they are recovered during environmental restoration of military installations.

In addition to evaluating commercial disposal as reported by the Army to Congress, this study reviews and evaluates the use of the RRS in two modes of deployment: the baseline mode, in which an RRS is transported to a site where CAIS are found or stored, and a "fixed RRS" mode, in which RRSs would be located at one or more fixed sites and CAIS from other sites would be shipped to one of these RRS sites for treatment. When the study committee learned that commercial facilities would probably dispose of CAIS by incineration, it decided to evaluate nonincineration alternatives to incineration that would be consistent with the congressional mandate for reducing program costs, as well as evaluating incineration at commercial facilities. Nonincineration methods would address the concerns of some public groups, including some stakeholders from communities near CAIS sites, that oppose the use of incineration for destruction of any chemical warfare materiel, including CAIS.

Because CAIS contain samples of chemicals considered to be chemical warfare agents, they are classified as chemical warfare materiel. For historical and legal reasons, the programs for destroying U.S. chemical warfare materiel are divided into two categories: programs for destroying eight specific stockpiles of materiel at sites in the continental United States and one site in the Pacific Ocean, and programs for destroying all other chemical warfare materiel either stored at military installations or recovered during environmental restoration. Federal law prohibits the facilities built to destroy the chemical stockpile materiel from being used to dispose of any other hazardous materials, including materiel in the non-stockpile category, such as CAIS. Other than the two modes of RRS operation, the study committee did not evaluate the alternative of disposing of CAIS at a facility built specifically for non-stockpile materiel because it does not address the congressional request. That option, which could provide another route for CAIS disposal, may be assessed as part of the committee's continuing review and evaluation of the Non-Stockpile Chemical Materiel Disposal Program.

The study committee evaluated the selected CAIS disposal options with respect to technology, laws and regulations, costs, environmental impacts, safety and health risks to workers and the public, involvement of a range of public and stakeholder groups, and programmatic considerations. The evaluations resulted in the conclusions and recommendations presented below.

CLASSIFICATION AND REGULATION OF CAIS FOR TRANSPORT AND DISPOSAL

The conclusions and recommendations on classification and regulation of CAIS apply to all of the disposal alternatives.

Conclusion 1. If existing Army policies and regulations, as well as U.S. laws and their interpretations, were clarified and made more internally consistent, CAIS disposal would be simplified and the number of disposal alternatives would be increased without compromising public safety. A consistent approach to regulating CAIS would be to classify the CAIS *set* or individual *items* from a set as a characteristic hazardous waste rather than as chemical warfare materiel or chemical agent. This approach is consistent with historical practice in environmental regulation, in which a waste is classified on the basis of the amount of chemical constituents it contains and the potential risks it poses. If CAIS sets and items were classified as a characteristic hazardous waste, this would not (and should not) set a precedent for reclassifying any of their chemical constituents, such as sulfur mustard, that are classified as chemical warfare agents or chemical warfare materiel when in other configurations.

Conclusion 1a. CAIS can be safely transported and handled if the best industrial practices for highly hazardous materials are used for packaging, handling, worker safety, monitoring, plant inspections, and audits, particularly if these practices are used in conjunction with the Army's experience in handling CAIS materials. Because either specialized commercial or Army-specific facilities and equipment could be used for transport and disposal, much of the present regulatory burden and Army bureaucracy surrounding the handling, transport, and disposal of CAIS items seems to be unnecessary.

Conclusion 1b. For the purposes of transportation and disposal, CAIS containing mustard and lewisite could safely be classified as hazardous waste and not as chemical warfare materiel. The reclassification would greatly reduce the costs of transportation and disposal and would substantially increase the feasibility of CAIS disposal. This change should have no impact on the safety of CAIS recovery, transportation, or disposal operations for the following reasons:

- CAIS contain no explosives.
- The chemicals in recovered or stored CAIS that are currently interpreted in Defense Department guidance as chemical warfare agents are sulfur mustard and lewisite. These chemicals are considered to have relatively high inherent hazard (at the high end of the range of hazards presented by hazardous industrial chemicals). Nevertheless, the risk posed by proper treatment of small quantities of these is less than the risk posed by the larger quantities of highly hazardous

industrial chemicals that are already handled by the chemical industry and commercial hazardous waste treatment facilities. Although some CAIS configurations contain potentially lethal quantities of chemicals, the risks to the public and workers in handling CAIS can be controlled to protect human health.
- Most CAIS (except for two types that contain several liters of agent per set) contain relatively small quantities of chemical ingredients, often in dilute forms.

Recommendation 1. The Army should present a plan to Congress describing how it will work with regulators, other appropriate decision makers, and stakeholders to clarify the regulatory status of Chemical Agent Identification Sets (CAIS), either through separate legislation (as part of 50 U.S.C. section 1512) or by other appropriate means. A range of stakeholders and public groups should be included in this process to ensure that this proposal to clarify regulations is presented in a forthright manner. In particular, the Army should inform the public that CAIS items contain chemical warfare agents and should be explicit about the technologies that would be used for commercial disposal. This plan should be part of the Army's overall program for CAIS disposal and should address ancillary issues, such as the implications of the Chemical Weapons Convention. One alternative that should be explored through this process is the feasibility of classifying complete CAIS sets or items from sets as a characteristic hazardous waste.

COMMERCIAL INCINERATION

Conclusion 2. Even though commercial incineration seems technically feasible and may offer cost and time savings compared to the RRS, many hurdles would have to be overcome. Not the least is ensuring that commercial incineration of CAIS is acceptable to the public.

Recommendation 2. If the Army and its stakeholders cannot agree that the commercial incineration of CAIS is practical, the Army should expand its inquiry to include other disposal alternatives, such as nonincineration disposal methods, in either Army or commercial facilities, using technologies that have already been used in operational, permitted facilities or are scheduled to be demonstrated.

Conclusion 3. It is technically feasible to dispose of all known CAIS items in commercial hazardous waste incineration facilities that have a permit specifically addressing wastes containing arsenic and that operate at the highest level of destruction and removal efficiencies for organic compounds. An example would be a permit specifying destruction and removal efficiencies similar to those required for commercial incineration facilities permitted to treat nitrogen mustard, polychlorinated biphenyls, or dioxins. Disposal in these commercial incineration facilities can be safe, reliable, and effective. The committee anticipates that a thorough and well-documented comparison of risk components will show that the risk to the public from the incineration of *smaller quantities* of CAIS items is lower than the risk from the routine incineration of *larger quantities* of highly toxic industrial chemicals. With appropriate process controls and monitoring, as discussed in this report, the committee also anticipates that risks to workers from incineration of CAIS items will be no greater than the risks from other commercially incinerated materials that are routinely handled in these facilities.

Recommendation 3. To provide a documented evaluation of the environmental and worker/public safety issues involved in the commercial incineration of CAIS, the Army should prepare a report that compares the relative risks to workers and the public of incinerating CAIS items with the risks to workers and the public of incinerating highly hazardous industrial chemicals at any facility proposed for CAIS disposal. Among the components of risk that should be documented are (1) the toxicity of chemical agents in CAIS (mustard and lewisite) relative to highly hazardous industrial chemicals (e.g., agent-contaminated materials, highly toxic industrial chemicals, polychlorinated biphenyls, medical wastes, and other hazardous military wastes) that are routinely destroyed in commercial incineration facilities; (2) the anticipated annual volumes of agents in CAIS to be disposed of, compared with the annual volumes of highly hazardous industrial chemicals that are currently being commercially incinerated; and (3) the Environmental Protection Agency's "incinerability" classifications of chemicals in CAIS and highly hazardous industrial chemicals.

Conclusion 4. By law, chemical warfare agent disposal facilities are required to provide maximum protection of the public, workers, and the environment. However, the term "maximally safe" is not clearly defined in the statute or in Army regulations and guidance documents.

Recommendation 4. Either the Army, the U.S. Department of Defense, or Congress should clarify the interpretation of "maximally safe" to ensure that it can be applied consistently in different situations. For the transportation and handling risks, the role of feasibility in determining what is maximally safe should be incorporated through the use of regulatory concepts such as ALARA (as low as reasonably achievable) or ALARP (as low as reasonably practicable). For the risks from emissions and discharges, the well-established regulatory policy for managing waste disposal risks should be applied. For all risks, a risk management approach should be used to ensure that appropriate controls are identified and evaluated.

Conclusion 5. The Army and its contractor conducted a preliminary analysis of the technical feasibility of commercial disposal of CAIS items at selected sites by incineration. The analysis was based on destruction of similar materials, and no trial burns were conducted. Sulfur mustard, the major chemical of concern in CAIS items, has been successfully destroyed via incineration and chemical neutralization. Lewisite, an arsenic-based material, has also been destroyed successfully, but, if it is incinerated, special scrubbing equipment may be required to meet regulatory limits on arsenic emissions. Although the committee does not know whether the facilities surveyed by the Army could handle arsenic-based materials, there are commercial incinerators that have permits allowing them to treat wastes containing arsenic. Characterization of incoming wastes (for compliance with a facility permit), monitoring of destruction removal efficiencies and emissions (particularly arsenic), and special handling (unless CAIS overpacks containing mustard or lewisite could be fed directly into the disposal equipment) may be required at commercial facilities. These requirements, combined with possible process and permit modifications, could be major economic and technical hurdles for commercial facilities.

Recommendation 5. The Army should develop a stronger technical basis for its conclusion that commercial incineration of items from Chemical Agent Identification

Sets (CAIS) is technically feasible (e.g., by determining if anything unique about CAIS disposal would preclude commercial incineration). The Army should also provide side-by-side data showing the destruction kinetics of CAIS and highly hazardous chemicals already being destroyed in commercial facilities. The data should be consistent with the conditions at state-of-the-art commercial facilities (i.e., facilities permitted to handle hazardous chemicals, such as polychlorinated biphenyls, dioxins, or nitrogen mustard).

Conclusion 6. A preliminary cost estimate developed by the Army and its contractor showed that commercial incineration of CAIS items could yield substantial cost savings compared with the RRS option. However, a number of items either were not included or were not adequately discussed in this preliminary cost estimate (e.g., permit modifications, transportation of CAIS items, packaging, agent monitoring and other facility modifications, and staff training). In contrast to this optimistic estimate, the projected costs of the Army's baseline approach (i.e., the mobile RRS) seem overly conservative. Furthermore, the preliminary estimate did not include programmatic issues for the commercial incineration option. If the commercial option is pursued, issues of corporate commitment, legal liability, public notification requirements, and contractual matters could arise.

The Army's cost estimate for commercial incineration was two orders of magnitude lower than the estimate for the RRS, which implies a potential for significant savings even after accounting for the costs not included in the estimate. However, given the potential regulatory problems, public concerns, and liability barriers, the Army may have to remove barriers before commercial firms will undertake CAIS disposal.

Recommendation 6. The committee concurs with the Army's finding that a comparative cost analysis of commercial facilities with the options for the Rapid Response System should be conducted. The existing analysis provided by the Army is inadequate for this purpose. The cost analysis should be more detailed and, to the extent possible, should include all relevant costs so that accurate comparisons can be made.

Conclusion 7. The Army's report to Congress did not include a risk assessment for the commercial (incineration) disposal option; in fact, it did not discuss the risks at all. However, because phosgene and other CAIS ingredients are routinely used and disposed of in the chemical industry in much larger quantities than occur in CAIS, it seems reasonable to assume that the risks during CAIS disposal could be controlled. The Army's risk evaluation framework for the Chemical Stockpile Disposal Program could be adapted for application to CAIS disposal options.

Recommendation 7. To characterize the risks of the commercial incineration option, the Army should conduct a risk evaluation using various hazard identification and evaluation methodologies, as appropriate. The evaluation of risks should include risks from delays, from transporting CAIS to a commercial facility, from handling CAIS in a commercial facility, and from treating CAIS disposal effluents. Worker safety during CAIS disposal should be evaluated using objective safety criteria to determine the degree of specialized personal protective gear, workplace monitoring equipment, and/or specialized training that may be necessary. If the evaluation indicates risks to workers or the public that appear to warrant further risk control measures, then more detailed risk assessments may be helpful. Commercial operations for CAIS disposal should use procedures and provide protection equivalent to the safety practices that have been determined to be necessary in military installations that handle CAIS.

Conclusion 8. The commercial incineration option may encounter public opposition by various groups, which could lead to schedule delays and added costs similar to those experienced by the Chemical Stockpile Disposal Program. Unfortunately, the Army's report to Congress did not include a detailed analysis of public acceptability issues—including how CAIS disposal would be related to the overall strategy for the disposal of non-stockpile materiel from the public's perspective. Instead, the report focused on cost, technical efficiency, and legal issues. Past experience has shown that focusing on these issues alone does not ensure public acceptability. Whichever option the Army favors, considerable staffing and funding for public involvement activities will be required to facilitate selection of an option that is both technically sound and acceptable to the public.

Recommendation 8. If the commercial disposal option is pursued, the Army should carefully assess the public acceptability challenges of commercial incineration and ensure that the necessary resources and staff (skills, experience, and number) are available to develop and implement an effective public involvement program. This program should be coordinated with similar activities Army-wide, particularly activities of the Chemical Stockpile Disposal Program, to ensure that the approaches to public involvement are consistent.

RAPID RESPONSE SYSTEM

The Army's current plans for CAIS disposal are based on the use of a transportable RRS, which is currently being tested. The committee found that the RRS, in both the baseline, mobile configuration and the fixed mode, offers advantages in mobility and simplicity of operation (important attributes from the public's perspective), as well as the capability to characterize, separate, and repackage individual CAIS items. However, the committee also found that operational costs, permitting requirements, and follow-on treatment of RRS wastes are issues that must be addressed prior to using either RRS configuration.

Mobile Rapid Response System

Conclusion 9. Although some national and regional stakeholder groups have endorsed the concept of a mobile facility, a number of unresolved issues will make the disposal of CAIS via the mobile RRS difficult. Preliminary cost estimates indicate that RRS deployments will be expensive and more time consuming than the Army originally envisioned. For example, state-by-state permit requirements will hinder the rapid use of the RRS, and processing and transport costs in the Army's estimate seem unusually high. The RRS neutralization scheme seems viable as a preliminary processing step, although the entire RRS has not yet been fully tested as a system, and issues surrounding the monitoring and subsequent disposal of process effluents, in particular the use of incineration for treating RRS wastes, have not been completely resolved.

Recommendation 9. As the Army begins initial testing of the Rapid Response System (RRS), it should critically examine a number of unresolved issues, including site-specific permitting requirements, monitoring, public involvement, and the disposal of process effluents. These issues should be resolved prior to the operational deployment of the RRS.

Conclusion 10. Only two sites have permits that would allow long-term storage of CAIS prior to the arrival of an RRS: Deseret Chemical Depot (Utah) and Pine Bluff Arsenal (Arkansas). Both sites have occasionally placed restrictions on the receipt of CAIS items. Regulatory approval for transporting CAIS items across state lines to these sites will also affect disposal costs and schedules. The procedural and regulatory requisites for transportation of CAIS could be simplified by preparing a generic plan or template with wording appropriate for all situations, such as descriptions of relevant regulations, the mode of transport to be used, handling procedures, and so on. This template could include blanks for situation-specific details, such as the locations from and to which CAIS are transported, the specific CAIS materials to be moved, and situation-specific risks to be addressed.

Recommendation 10. The Project Manager for Non-Stockpile Chemical Materiel should work with the Deseret Chemical Depot (Utah) and Pine Bluff Arsenal (Arkansas) storage facilities to clarify their acceptance criteria for Chemical Agent Identification Sets or items from them. The project manager should also consider developing alternative storage facilities in case these facilities become temporarily unavailable. The Army should work with regulators to reduce the time and administrative costs of developing transportation plans, recognizing that portions of these plans will necessarily be site-specific.

Fixed Rapid Response System

Conclusion 11. Disposal of CAIS by means of the fixed RRS approach seems to offer potential cost savings by reducing the requirements for site-specific disposal permits and facility transportation. However, transporting CAIS to a fixed RRS will require regulatory approval and may be less attractive to some members of the public than a mobile facility. Regulatory costs could be significant unless the Army can obtain generic transportation permits or other forms of administrative relief.

Recommendation 11. If the fixed (regional) option for the Rapid Response System is pursued, the Army must move quickly to engage base commanders, regulators, and public and stakeholder groups in exploring the details of this approach, including the disposal of process effluents and the locations of the fixed facilities.

NONINCINERATION-BASED OPTIONS

Conclusion 12. The Army's Alternative Technologies and Approaches Program and Assembled Chemical Weapons Assessment Program have identified several nonincineration technologies for the disposal of chemical warfare agents, including sulfur mustard and possibly lewisite. These processes may be more acceptable to the public than either commercial incineration or neutralization of CAIS materiel in the RRS followed by incineration. Nonincineration processes might be implemented in either commercial or

Army-owned facilities. However, the absence of economic incentives for commercial firms to make process and regulatory modifications may preclude the use of commercial facilities.

Recommendation 12. The Army should evaluate the technical feasibility of using nonincineration processes for destroying Chemical Agent Identification Sets and process effluents. The Army should also consider methods of identifying and overcoming institutional, regulatory, and economic barriers to the development of commercial nonincineration facilities.

Conclusion 13. The disposal of CAIS in Army facilities that use nonincineration methods of destruction could offer a low-cost, maximally safe option, if CAIS disposal can be conducted as part of the normal, planned operations of these facilities. The technology being used by the Chemical Agent Munitions Disposal System may be appropriate for the disposal of CAIS items containing lewisite. The neutralization-based technology planned for the facility at Aberdeen Proving Ground may be appropriate for the disposal of CAIS items containing mustard. The Army has explicitly promised concerned stakeholders not to seek to alter the federal law prohibiting the use of chemical stockpile disposal facilities for the disposal of other wastes, including CAIS. Therefore, public resistance and current legal restrictions on additional uses of the stockpile facilities may make their use for CAIS disposal impossible. Nevertheless, the use of nonincineration-based disposal technologies like those at existing or planned Army facilities appears to be a technically and economically attractive option for the disposal of CAIS containing mustard or lewisite, provided affected communities agree and are involved in establishing the conditions for use of the facilities.

Recommendation 13. Congress should consider revising the legal restrictions on the use of stockpile disposal facilities to allow the disposal of Chemical Agent Identification Sets (CAIS) at appropriate nonincineration-based facilities, at least where the local community agrees to short-term use of a facility to dispose of limited amounts of recovered and stored CAIS materials. At the same time, the Army should explore the use of nonincineration-based technologies for CAIS disposal and should engage the affected public and stakeholders at sites that will use these technologies in exploring the acceptability of this alternative.

A PATH FORWARD

The committee found that, if legal and regulatory burdens can be reduced, the CAIS disposal program could be accelerated safely, reliably, and effectively. However, implementation would require changes in current law and policy, with the advice and consent of the public.

Although the committee believes that incineration of CAIS under controlled conditions is technically acceptable, some members of the public have expressed strong opposition to incineration. Based on experience with other disposal programs and the committee's interactions with concerned public groups, the committee expects that the public may be more accepting of disposal technologies that are not based on incineration.

Summary Conclusion. All of the alternatives for disposing of CAIS evaluated by the committee have advantages and disadvantages. Although the approach, or approaches, will ultimately be selected by the Army, the committee believes the Army can take several steps to expand its options. As the Army moves forward, it will be vital that a range of public and other stakeholder groups be actively involved in decision making. The committee believes that consideration of the perspectives of these groups on risk, economic implications, and other aspects of CAIS disposal options will contribute significantly to better decisions.

Summary Recommendation. The Army should take the following actions to expand its options for cost-effective disposal of Chemical Agent Identification Sets (CAIS) without decreasing safety or increasing the risks to workers, the public, or the environment:

- The Army should reconsider its interpretation of CAIS as chemical warfare materiel under U.S.C. section 1512. If the Army decides it cannot change its interpretation, then Congress should consider amending the legislation so that CAIS sets or items from CAIS can be regulated as hazardous waste under the Resource Conservation and Recovery Act.
- The Army should promote the development of nonincineration technologies for CAIS disposal.
- The Army should develop, review with stakeholders, and then implement a written plan for public involvement designed to reach a range of stakeholders and concerned groups, including affected communities and tribal nations, state and federal regulators, concerned national and regional groups, and representatives of the waste disposal industry.
- In states with a chemical stockpile disposal facility, the Army should engage the affected communities in a discussion of alternatives, including the potential use of the stockpile facility for CAIS disposal. If a community agrees to consider using the stockpile facility (and only if it agrees), the Army should pursue that option with the full involvement of the community, including establishing specific conditions for the use of the facility. If the community agrees, which may be more feasible at facilities that use nonincineration technologies, the law prohibiting the use of chemical stockpile disposal facilities for any other purpose would have to be modified to allow CAIS disposal.
- An important current capability of the RRS is that it can characterize, separate, and repackage individual CAIS items. However, because of the inherent permitting problems and high costs of the mobile RRS option, the Army should aggressively pursue other options while continuing to implement the RRS.

1

Introduction

The U.S. Army is preparing to dispose of thousands of chemical agent identification sets (CAIS), test kits used from 1928 to 1969 to train solders in defensive actions in case they came under chemical attack. A typical CAIS set includes small glass vials (ampoules) or bottles containing various chemical warfare agents or hazardous industrial chemicals, which could be, or have been, used in chemical warfare.[1] Many of the CAIS to be destroyed are in storage on Army bases, but CAIS sets and individual CAIS items are also being recovered during cleanup operations at current or former military installations.[2] Some CAIS sets or items have also been discovered by the public on sites of former military installations.

For reasons related to the history of international treaties and U.S. law concerning the disposal of chemical warfare agents and the munitions or components associated with their use, the totality of chemical warfare materiel under U.S. control falls into two categories. One category consists of a well-defined "stockpile" of chemical agents and related munitions and other materiel. The other category includes everything else. Thus, under the Army Program Manager for Chemical Demilitarization, which has overall responsibility for disposing of both categories of materiel, there is a Chemical Stockpile Disposal Program and a Non-Stockpile Chemical Materiel Program (NSCMP). CAIS are included in the non-stockpile category, so the NSCMP is responsible for their disposal. (For a more detailed description of the NSCMP, see Appendix C, as well as the introductory discussion later in this chapter.) Public Law 99-145, which defined the stockpile, also prohibits the use of stockpile disposal facilities for any other purpose, including the disposal of CAIS or other non-stockpile chemical warfare materiel.

In 1997, Congress directed the Secretary of Defense to assess the policy and plans for disposing of CAIS and to report specifically on disposal alternatives and changes in policy that "could result in significant reductions in the cost of the non-stockpile program with no reduction in overall program safety." The task of preparing the assessment report was assigned to the NSCMP, and its report to Congress was delivered in August 1998 (U.S. Army, 1998a).

---

[1] For the purposes of this report, "industrial chemicals" includes the wide range of chemicals used by industry for civilian (nonmilitary) purposes. In many cases, these chemicals are classified as "hazardous waste" by the Environmental Protection Agency (EPA) when they are disposed of. If industrial chemicals classified as "hazardous wastes" are accidentally spilled, the soil containing the chemicals may be considered a "hazardous substance," as that term is defined in the Comprehensive Environmental Response Compensation and Liability Act (CERCLA, commonly known as Superfund).

[2] To avoid confusion in this report, the term "CAIS set" refers to a complete kit, "CAIS item" refers to recovered individual bottles or ampoules, and "CAIS chemical" refers to the agent or industrial chemical contained therein.

STATEMENT OF TASK AND CONGRESSIONAL DIRECTION

In the fall of 1997, the Army requested that the National Research Council (NRC) review and evaluate the NSCMP. The NRC Governing Board approved the formation of the Committee on Review and Evaluation of the Army Non-Stockpile Chemical Materiel Disposal Program ("the committee"). During its first year, the committee's tasks were to become familiar with the program and evaluate the Army's report to Congress. During its second and third years, the committee's task will be to recommend improvements in the overall plan for the NSCMP. The Governing Board approved the following Statement of Task on March 10, 1998, as the charge to the committee:

> The NRC will:
>
> Develop a comprehensive understanding and knowledge base concerning the destruction of non-stockpile chemical warfare materiel (CWM) and provide recommendations in two written reports. These reports will (1) evaluate a DoD assessment of its policy on disposal of chemical agent identification sets (CAIS); (2) suggest changes to the mid-term that may lead to significant improvements. The working knowledge to be achieved by the NRC will concentrate on the mission, philosophy, objectives, and methodology of the non-stockpile project, which includes: current baseline systems, transportation and storage issues, monitoring and standards, environmental laws and regulations, public outreach and community involvement, and applicable technologies.
>
> Specific NRC reporting requirements are to:
>
> Prepare a first report that evaluates the Department of Defense assessment of policy aspects controlling chemical demilitarization of chemical agent identification sets (CAIS). DoD will report to Congress by March 31, 1998, on its assessment and on coordination efforts with the NRC. The NRC will examine and comment on the DoD assessment in the form of an NRC report by February 28, 1999.
>
> Examine the non-stockpile project for the mid-term. This examination may address the following: assessment and access of munitions, monitoring and standards development, agent and explosive containment, treatment of neat agent, post-treatment processes (including waste disposal), and public stakeholder concerns. The Committee will issue an NRC report by January 31, 2000, which will make recommendations regarding the conduct of the project.
>
> Each report will address any NRC interactions with public stakeholders conducted to ascertain public acceptance of the disposal technologies under consideration.

This report addresses the first of the two reporting requirements. After the Army's report to Congress was delivered in August 1998, the completion date for this evaluation was extended to October 1999. The committee has begun working on the second part of its task, which will culminate in a second report sometime in 2001.

The Congressional Mandate

The committee read the Army's report to Congress on CAIS disposal in light of the original congressional mandate to which the NSCMP was responding. The mandate was included in Conference Report 105-340 (dated October 23, 1997), which accompanied the National Defense Authorization Act for Fiscal Year 1998 (Public Law 105-85):

> The conferees understand that a major aspect of the chemical non-stockpile materiel project is the development of a system for disposal of the chemical agent identification kits, which have been classified as chemical weapons/agents for the purpose of the chemical disposal program, rather than hazardous waste. The conferees direct the Secretary of Defense to conduct an

assessment of its policy, which includes chemical agent identification kits in the chemical agent demilitarization program, the current plans for disposal, and the potential changes in policy and disposal alternatives that could result in significant reductions in the cost of the non-stockpile program with no reduction in overall program safety. The assessment shall be conducted in coordination with the National Research Council. The results of the assessment and the Secretary's decision should be provided to the congressional defense committees by March 31, 1998.

Initial Interpretation and Evolution of the Task

The committee interpreted the congressional direction as stipulating two constraints to the scope of its first-year reporting task. First, the committee's report should be an evaluation of the Army's report, not a separate exploration of a broad range of technical possibilities unrelated to what Congress specifically requested from the Secretary of Defense. Second, the disposal options considered should be consistent with the congressional interest in "potential changes in policy and disposal alternatives that could result in significant reductions in the cost of the non-stockpile program with no reduction in overall program safety." As explained in Chapter 4, the Army's report to Congress focused on the use of commercial disposal facilities for CAIS disposal as the option that could most significantly reduce cost without reducing overall safety. Therefore, a major focus of this report is the Army's proposed plan for using commercial facilities to dispose of CAIS. The Army's "current plans for disposal" at the time of the congressional request depended primarily on the Rapid Recovery System (RRS), a transportable processing facility that neutralizes CAIS chemicals in a small chemical reactor. Thus, the RRS, as the "baseline" system for CAIS disposal against which commercial disposal was evaluated, is another focus of this report.

When the committee began to investigate the commercial option for CAIS disposal, it learned that commercial facilities with the appropriate permits and technology would probably use incineration to dispose of CAIS. In exploratory interviews by committee staff with some of the commercial firms surveyed by the Army, the firms said that because of the reliability of incineration and "simple economics," they would use incineration for the disposal of CAIS, even if nonincineration-based disposal technologies were available. The Army's report to Congress did not specify the disposal technologies that would be used by a commercial facility to dispose of CAIS.

Given the long history of public reaction to the incineration-based disposal of the stockpile materiel (see Public/Stakeholder Involvement in Chapter 3), the committee decided that the issue of whether commercial disposal would involve incineration or nonincineration technologies could significantly affect the feasibility of a commercial disposal alternative to the RRS. Representative nonincineration technologies that are, or reasonably could be, employed by a commercial treatment, storage, and disposal facility (TSDF) were therefore included in this report. Finally, a potentially low-cost option that may satisfy stringent technical criteria is the use of nonincineration facilities designed for the disposal of stockpile materiel. However, this option raises significant issues of federal law, Army policy, and commitments made by both Congress and the Army to affected stakeholders.

Two deployment options for the RRS are evaluated in this report: the baseline approach, in which the RRS is transported to each site where CAIS are stored or found, and a fixed RRS mode, in which RRSs are located at one or more sites, to which CAIS from other sites are transported. In addition to the RRS, which is an Army facility dedicated to CAIS disposal, a technologically feasible alternative would be an Army facility designed for disposal of non-stockpile materiel in general, including CAIS. The committee did not evaluate that option for this report because it goes beyond the congressional mandate to address CAIS disposal. The general disposal of non-stockpile materiel will be the subject of the second and third reports from the committee

Report Structure

Chapter 1 contains introductory information on chemical warfare materiel in general and CAIS in particular. This background information is important for understanding the Army's report to Congress and the committee's evaluation of that report. The CAIS disposal alternatives that the committee considered, in addition to the Army's commercial disposal option, are presented in Chapter 2. The evaluation methodology the committee applied to the disposal alternatives is discussed in Chapter 3. In Chapter 4, commercial disposal, as described in the Army's report to Congress, is evaluated according to this methodology. In Chapter 5, the other disposal alternatives selected by the committee are evaluated. Chapter 6 contains the conclusions and recommendations the committee drew from its analyses and evaluations. The laws and regulations that control CAIS disposal alternatives are explained in Appendix D.

CHEMICAL AGENTS AND CAIS

Chemical weapons were first used on the battlefield during World War I, when the Germans released chlorine gas in 1915. Chlorine and other choking agents, such as phosgene, burned the lungs and caused panic among soldiers who were unprepared for them. As gas masks became more effective against inhaled poisonous gases and were more widely deployed on the battlefield, blister agents such as mustard were employed (e.g., at Ypres, Belgium, in 1917). (Nerve agents, such as GB and VX, and blister agents are the principal constituents of chemical weapons that still exist around the world.)

Types of Agents Found in CAIS

In general, chemicals that have been used for military purposes to incapacitate opposing personnel (chemical warfare agents) can be classified as nerve, blister, vomiting, choking, or riot control agents. CAIS were manufactured with samples of various types of these chemicals, but not all chemical warfare agents in the U.S. inventory were included in CAIS. Table 1-1 lists the names and formulas of chemical warfare agents and agent-simulants that were included in one or more types of CAIS. Table 1-2 contains information on the physical properties and toxicity for these CAIS chemicals. This section briefly describes the broad categories of chemical warfare agents but focuses on the agents in CAIS.

GB is the only nerve agent that was ever included in CAIS. Like other nerve agents (e.g., VX, tabun, and soman), GB functions by inhibiting acetylcholinesterase, which causes an accumulation of the neurotransmitter acetylcholine at nerve endings. The nerve fibers are overstimulated, causing uncontrolled and disorganized of the functioning organs. Typical effects of this phenomenon include excessive secretions of saliva and tears, muscle twitches, jerky random movements, disorientation, and convulsions. Severe exposures can lead to coma and death from respiratory paralysis. Very low levels of GB are toxic, and fatal quantities can readily be absorbed through the skin or respiratory tract.

TABLE 1-1 Chemical Names and Formulas of CAIS Chemicals

Symbol	Common Name	Chemical Name	Formula
CG	phosgene	carbonyl chloride	$COCl_2$
CG-sim	triphosgene (phosgene simulant)	hexachloromethylcarbonate	$(OCCl_3)_2CO$
CK	cyanogen chloride	Chlorine cyanide	$CNCl$
CN	chloroacetophenone	Phenyl chloromethyl ketone	$C_6H_5COCH_2Cl$
DM	adamsite	Diphenylamine chloroarsine	$C_6H_4(AsCl)(NH)C_6H_4$
GA-sim	ethyl malonate	Diethyl malonate	$CH_2(COOC_2H_5)_2$
GB	sarin	Isopropyl methyl phosphonofluoridate	$(CH_3)_2CHO(CH_3)PFO$
H	sulfur mustard	HD plus impurities	HD plus impurities
HD	sulfur mustard distilled	2,2'-dichlorodiethyl sulfide	$S(CH_2CH_2Cl)_2$
HS	sulfur mustard in solvent	H or HD with 15% diluent	H/HD with 15% CCl_4
HN	nitrogen mustard	See HN1 or HN3	See HN1 or HN3
HN1	nitrogen mustard	2,2'-dichlorotriethylamine	$N(CH_2CH_2Cl)_2(C_2H_5)$
HN3	nitrogen mustard	2,2',2"-trichloroethylamine	$N(CH_2CH_2Cl)_3$
L (or M-1)	lewisite	2-chlorovinyldichloroarsine	$ClCH:CH-AsCl_2$
PS	chloropicrin	nitrotrichloromethane	CCl_3NO_2

Sulfur mustard (e.g., HD) and lewisite are blister agents, or vesicants.[3] Blistering compounds are readily absorbed through skin in liquid or vapor form and are distributed systemically. Like GB, low levels of HD are highly toxic, and lethal quantities can readily be absorbed through the skin. The acute lethal dermal dose of HD is slightly lower than that of GB; the acute lethal inhaled dose is somewhat higher than that of GB. Lewisite an arsenic-containing blister agent, is considerably less lethal than the sulfur mustard compounds, although its blistering action is equivalent to that of HD. Both HD and lewisite are carcinogenic.

Adamsite is a vomiting agent. Exposure causes pain in the nose and throat, severe headache, and violent, uncontrollable sneezing, coughing, nausea, and vomiting. The effects are delayed by several minutes and can last up to two hours.

Phosgene and chloropicrin are generally classified as choking or lung-damaging agents because of their severe irritant action in the lungs. At sufficiently high concentrations, they can cause pulmonary edema (buildup of fluid in the lungs), which is usually delayed but can be fatal. Other choking agents include chlorine and diphosgene. The immediate effects of exposure to phosgene include mild irritation of the eyes and lungs, but these early symptoms can be misleading because severe, life-threatening pulmonary edema can occur hours later.

---

[3]Although nitrogen mustard (HN) is also a powerful blistering agent, its propensity to inhibit cell division has made it useful for the treatment of cancer, and it is therefore not classified as a chemical warfare agent. Lewisite was developed in 1918 by W. Lee Lewis while Lewis was working at the chemical laboratory at the Catholic University of America in Washington, D.C. Limited quantities were produced during World War I, but lewisite was not used on the battlefield.

TABLE 1-2 Characteristics and Biological Effects of CAIS Chemicals

Agent	Appearance	Odor	Effect on Body	Rate of Action
Adamsite (DM)	Yellow to green solid	None in pure form	Vapor causes severe pepper-like irritation of the nose, throat, and eyes; tearing; severe headache; acute pain and tightness in the chest; and violent, uncontrollable sneezing, coughing, nausea, and vomiting.	Several minutes
Chloroaceto-phenone (CN) (tear gas)	Colorless to gray solid	Apple blossoms	Vapor causes headache, irritation, burning, and pain of the nose and throat and copious tearing. Dermal exposure causes transient burning, itching, and blisters on tender areas of the skin. Minimal irritant concentration is 0.3 mg/m^3; a concentration of 350 mg/m^3 may be life threatening.	Immediate
Chloropicrin (PS)	Colorless, oily liquid	Stinging and pungent	Vapor causes severe skin, eye, nose, and throat irritation with tearing, coughing, nausea, and vomiting. Pulmonary edema may develop hours after exposure to high concentrations. Contact with liquid causes severe skin burns and blisters.	Immediate
Cyanogen chloride (CK)	Colorless liquid (evaporates quickly)	Pungent biting odor	Vapor causes intense irritation of the eyes, nose, and throat, with coughing, tightness in the chest, tearing, difficulty breathing, and possible pulmonary edema. Effects of moderate exposure include dizziness, vomiting, and incontinence. Severe exposures cause convulsions and coma. Contact with liquid burns the skin and eyes.	Immediate
GA simulant (diethyl malonate)	Colorless liquid	Sweet ester odor	Vapors irritate the respiratory tract. Ingestion causes sore throat, abdominal pain, diarrhea, eye and skin irritation with redness and pain	
Lewisite (L)	Colorless to amber liquid	Geraniums	Vapor causes severe, painful eye and skin irritation immediately after exposure, followed hours later by blistering. Inhalation causes severe irritation of the respiratory tract with burning sensation, coughing, and sneezing; pulmonary edema may develop hours after exposure.	Immediate

TABLE 1-2 continued

Agent	Appearance	Odor	Effect on Body	Rate of Action
Nitrogen mustard (HN)	Oily, colorless to pale yellow liquid	Fishy or musty odor	Low vapor concentrations can cause eye irritation within 20 minutes of exposure in the absence of other effects. Higher vapor concentrations or exposure to liquid causes itching and reddening of the skin followed by blistering. Inhalation irritates the respiratory tract causing hoarseness, loss of voice, and persistent coughing. A fatal bronchopneumonia may develop after 24 hours. Ingestion causes nausea and vomiting, bloody diarrhea, intestinal lesions, and may depress normal formation of blood cells.	Within 20 minutes
Phosgene (CG)	Colorless gas	New-mown hay	Immediate responses to concentrations as low as 2–5 ppm include tearing, burning of the nose and upper respiratory tract, coughing, vomiting, chest tightness, and difficulty breathing; 50 ppm may be rapidly fatal.	Immediate
Sarin (GB)	Colorless liquid	None in pure form	Exposure can cause runny nose, excessive salivation and tearing, miosis (constricted pupils), dim vision, eye pain, tightness in the chest, difficulty breathing, and muscle twitches. Higher concentrations cause nausea, vomiting, abdominal cramps, incontinence, convulsions, coma, respiratory failure, and death.	Minutes for vapor exposure; 2 hours for dermal exposure
Sulfur mustard (HD)	Colorless to pale yellow liquid	Garlic	Early symptoms include severe irritation of the eyes, skin, and respiratory tract; vomiting, fever, and redness of the skin (erythema); severe blistering and ulceration of exposed areas may develop 4–24 hours later. Lethal quantities may be absorbed through the skin or lungs.	Usually 4 to 6 hours, but may be 24 hours depending on concentration
Triphosgene (phosgene simulant)	White to off-white crystals		Exposure to vapor can cause irritation of eyes, respiratory tract, and skin. Fatal pulmonary inflammation and edema may develop hours later.	Within 10 minutes to 2 hours, depending on concentration

Sources: Proctor and Hughes, 1978; NATO, 1995; U.S. Army, 1995a.

> BOX 1-1 Use of CAIS
>
> CAIS were originally intended for training combat troops in the identification of the smell and effects of chemical agents. Soldiers were intentionally exposed to the chemicals in the CAIS to enable them to recognize the odor and effects of chemical agents and to train them to take immediate defensive action. According to a training manual, "every soldier should become proficient in identification of gases through odor and other sensory reactions, since other means may not be available" (War Department, 1944).
>
> The recommended method of training was to detonate the glass vials with blasting caps to atomize the chemicals and form a small aerosol "cloud." The trainees were positioned downwind prior to the detonation and were instructed to allow the cloud to envelop them (or walk into the cloud if the wind had blown it away from them), "take a sniff, just enough to recognize the odor," and "to walk out of the cloud to the flank and exhale."
>
> According to the manual, "Normally, four gases are detonated in succession, with an interval between gases. For effective instruction the name of the gas should not be announced before it is fired." The trainees were graded on their ability to identify the four gases in order. "Men who fail to identify the gases should go through the exercise again. It should be made clear to them that this is an opportunity, not a penalty, for their lives may later depend on their individual judgement."
>
> It is difficult to estimate the number of trainees who were exposed to this training. A conservative estimate can be made by multiplying the number of CAIS believed to have been used during training (assume 80,000 of the approximately 90,000 not destroyed at Rocky Mountain Arsenal) by the number of vials in each CAIS (24–48) for a total of 1,920,000 to 3,840,000 vials. If two vials of the four agents were used in each training exercise (a total of eight vials per session), as described in the training manual, between 120,000 and 240,000 sessions (exposure of a class of trainees to all four chemicals) were conducted. Conservatively assuming that each training class consisted of only five trainees, the estimated total number of trainees exposed to CAIS chemicals is 600,000 to 1,200,000.
>
> [1]War Department. 1944. Use of Chemical Agents and Munitions in Training. T.M. 3-305. Washington, D.C.

Chloroacetophenone, commonly known as tear gas, is classified as a riot control agent. These agents are typically characterized by low toxicity and rapid onset of effects, which are transient and of short duration. Exposure to chloroacetophenone causes immediate, but transient, burning, pain, and tearing of the eye and severe irritation of the respiratory tract. In recent years, chloroacetophenone has largely been replaced by less toxic compounds.

Chemical Agent Identification Sets

CAIS sets and CAIS items are the most commonly recovered kinds of buried non-stockpile materiel. Between 1928 and 1969, the Army used several types of CAIS to train soldiers and sailors to identify chemical warfare agents, typically by using "sniff sets" during classroom training (Box 1-1). In some instances, CAIS chemicals were vaporized in a controlled detonation; trainees then walked into the vapor cloud, sniffed the gas, and identified the agent based on odor (War Department, 1944).

INTRODUCTION

FIGURE 1-1 Army photographs of four CAIS types: (a) a toxic gas set, containing bottles of sulfur mustard for use in decontamination training exercises; (b) war gas identification set containing ampoules of neat or dissolved agents and simulants for use in outdoor training; (c) an instructional "sniff set" with agent-impregnated charcoal for use in classroom-based indoor training for agent identification; and (d) the Navy "sniff set" for classroom-based identification training. Sources: U.S. Army, 1993, p. 5-7; U.S. Army, 1995a, p. 5-10 and Appendix E.

CAIS were produced in large quantities, were widely distributed, and came in more than a dozen different types, grouped into three major varieties (see Figure 1-1 and Table 1-3). Toxic gas sets (sets containing two dozen or more glass bottles, each with about 100 ml of neat agent) were used for training in decontamination. War gas identification sets (small ampoules of neat agent or simulant, or agent in a solvent) were used for outdoor training. The sniff sets (agent or simulant on charcoal) were used for indoor classroom training.

CAIS of the same major variety have similar markings and packaging. As manufactured, they generally contained a few dozen glass ampoules or bottles of chemical agent or simulant packed in metal shipping containers or wooden boxes. The chemicals were either neat, in 5 percent solutions in chloroform, or adsorbed on granular activated charcoal. In the CAIS types that are still known to exist, the only contents classified as chemical warfare agents are HD and lewisite (Fatz, 1997). (As explained in footnote *a* to Table 1-3, all the CAIS that contained GB were reported to have been destroyed.) The

TABLE 1-3 CAIS Types and Components[a]

Type and Use	Number and Type of Container	Chemical Agents	Agent per Container (milliliters)	Agent per Set (liters)
K941 (toxic gas set, M1).[b] Used from WWII to the late 1950s for training in decontamination of vehicles or equipment while in protective clothing.	24 4-ounce round glass bottles	neat H, HS, or HD	103.3	2.48
K942 (toxic gas set, M2).[c] Used during the Korean War period for training in decontamination.	28 glass heat-sealed ampoules	neat H, HS, or HD	112.5	3.15
K951/K952 (war gas identification set, M1).[d] Used from the early 1930s to late 1950s for identification of agents using detector kits.	48 pyrex heat-sealed ampoules	12 ampoules of H in 38 ml of chloroform	2	0.024 H; 0.024 L; 0.24 PS; 0.48 CG
		12 ampoules of L in 38 ml chloroform	2	
		12 ampoules of PS in 20 ml chloroform	20	
		12 ampoules of neat CG	40	
K953/K954 (war gas identification set, AN-M1A1).[e] Used during the Korean War period for identification of agents using detector kits.	48 ampoules	8 ampoules of H in 38 ml chloroform	2	0.016
		8 ampoules of HN in 36 ml chloroform	4	0.032
		8 ampoules of L in 38 ml chloroform	2	0.016
		8 ampoules of neat CG	40	0.32
		8 ampoules of neat CK	40	0.32
		8 ampoules of neat GA-sim	40	0.32
K955 (Navy/sniff set, M1).[f] Used from the late 1930s to World War II for classroom-based training in identification by odor.	7 4-ounce round glass bottles	2 bottles of HS on 90 cc charcoal	25	0.05
		1 bottle of L (or M-1) on 90 cc charcoal	25	0.025
		1 bottle of PS on 90 cc charcoal	25	0.025

TABLE 1-3 Continued

Type and Use	Number and Type of Container	Chemical Agents	Agent per Container (milliliters)	Agent per Set (liters)
		1 bottle of neat CG-sim	6 grams	6 grams
		1 bottle of neat CN	15 grams	15 grams
		1 bottle of neat DM	15 grams	15 grams

^aApproximately 2,000 K945 (Chemical Agent Identification and Training Set [CAITS], M72 [Nerve Agent Sets]) were also produced and used until the late 1960s, but all are believed to have been destroyed. These sets consisted of eight agent bottles (four bottles of 3 ml GB nerve agent, one bottle of 3 ml of lewisite, one bottle of 1 ml of H, all on plastic pellets; and one bottle each of 5 grams of triphosgene and potassium cyanide). The bottles were contained in a plastic "tackle box" housed in a small wooden box. The M72A1 is a similar looking nerve-simulant set.

^bCommonly called "bulk mustard set." Sets of four bottles (labeled with heat-resistant paint) were packed with sawdust in a pressure-sealed, metal, 6.5" high can with a sardine-type key on the bottom. Six of these labeled cans were placed in a 38" long steel shipping cylinder (a PIG) with a bolted, flanged end cover. This type of CAIS is found either as complete PIGs or as loose bottles. When found loose, the plastic/bakelite screw-top bottle tops tend to leak because mustard is a good solvent for plastic and rubber. (However, mustard forms a scale or sludge in contact with soil or sawdust and solidifies at cool temperatures.)

^cCommonly called "bulk mustard set." Each 1 7/8" diameter by 4 5/8" long ampoule was packed in its own sealed metal can, surrounded by corrugated fiberboard lining and foam rubber to prevent breakage. The 28 gray cans (with green stenciled markings) were placed in steel bolted, flanged steel drums.

^dEach 1" diameter by 7 ½" long ampoule was packed in a screw top labeled cardboard tube. Twelve tubes were packaged into press fit metal cans, with four cans per 38" long steel PIG with a bolted, flanged end cover. The K951 was issued with blasting caps that were packed and shipped in a separate container.

^eTwelve ampoules were packed per can; four cans per PIG. The K953/K954 were the later versions of the K951/K952 War Gas Identification Sets, incorporating nitrogen mustard, cyanogen chloride, and ethyl malonate (GA nerve agent simulant).

^fBottles had etched labels. Each bottle had either a screw top or stopper, which was usually wax coated. Individual bottles were housed in 4" diameter by 7" high metal cans with a paint-can type lid. The complete K955 set was packed in a hinged, wooden footlocker type box, but K955 bottles have frequently been found loose at burial sites. The Navy Replacement Sets (X302, X545-X552 series) were also produced and used by the Navy to replace components of the K955. These sets contained the same types and amounts of agent as the K955 but included the nitrogen mustards.

Source: U.S. Army, 1995a; Brankowitz, 1998; and Chemical Warfare School training materials provided by committee member James Pastorick.

other chemicals found in existing CAIS (e.g., phosgene and chloropicrin) are classified by the Army as hazardous industrial chemicals. CAIS do not contain explosives or other energetics, such as the bursters, fuses, and propellants found in assembled chemical weapons.

Recovered CAIS

Approximately 110,000 CAIS were produced between 1928 and 1969. In 1969 the use of CAIS was discontinued.[4] All unused CAIS in Army stocks at that time, including all CAIS containing the nerve agent GB, were sent to Rocky Mountain Arsenal, where 21,458 of them were destroyed by incineration between 1979 and 1982 (Brankowitz et al., 1983; U.S. Army, 1998a). Most of the 90,000 CAIS not destroyed in this way are thought to have been used during training, with some or all of their chemicals expended. The exact

---

[4]Training with live agent was not conducted after the late 1960s, when the military began using simulants for training. For example, when one committee member went through training, antifreeze was used to simulate agents, such as VX, not for agent (smell) identification but rather to set off portable agent monitoring alarms. The alarm kits (and other newer and more sophisticated tests) can be easily fooled by some common substances, which give false positive indications for agent. These substances are used as simulants during training in the use of the alarm kits and in field operations in agent-contaminated environments. Reportedly, chlorine chambers are still used for training in the use of gas masks.

numbers of used and buried CAIS are not known because no detailed records were maintained. An unknown number of CAIS were disposed of by burial, either as sets in their wooden or metal containers (package in-transit gas shipment, or PIG, containers) or as loose CAIS items (ampoules or bottles). Some CAIS chemicals were disposed of as the process effluents from simple neutralization or burning.[5] These disposal procedures, including burial of CAIS sets or items, were standard and approved at the time.

Today, CAIS sets and items are being found at former World War II and Korean War training sites located at active or former military installations. Environmental restoration programs at former military installations are recovering additional CAIS. The identification of the chemical content of individual CAIS items may require sophisticated characterization equipment. PIG containers and other packaging, even if intact, are no longer considered safe for transporting CAIS materials off the site where they are found. Therefore, either the entire CAIS must be enclosed in a permitted container or the individual items must be removed by properly trained and protected personnel and transferred to authorized laboratory-type overpacks for transport.

After arrival at a disposal facility, entire sets would have to be removed from the transport overpacking. The original set containers, if intact, would presumably be opened before disposal. Individual items in laboratory overpacks, if characterized prior to transport, might be incinerable without unpacking. Disposal by a nonincineration process would probably require opening and emptying the CAIS items to ensure efficient destruction.

Approximately 10,000 CAIS items and 1,400 non-stockpile chemical munitions are currently in storage awaiting destruction (Table 1-4). The Army's report to Congress (U.S. Army, 1998a, Appendix D) listed 33 sites where CAIS items are, or may be, buried (Table 1-5). At the time of that report, near-term (fiscal year 1998–1999) CAIS recovery efforts were anticipated at Gerstle River Expansion Area, Alaska; Fort McClellan, Alabama; the former Santa Rosa Airfield, California; England Air Force Base, Louisiana; the former Plattsburgh Air Force Base, New York; the former Defense Depot, Memphis, Tennessee; and Ogden Depot, Utah.[6]

The Army is developing the RRS to (1) access CAIS items (i.e., open the CAIS packaging and then open the items containing the chemicals), (2) repackage the contents classified as industrial chemicals, and (3) chemically neutralize any chemical contents classified as chemical warfare agents (sulfur mustard or lewisite). Under current plans (fiscal year 1999), the waste products from RRS operations and the repackaged industrial chemicals will then be treated and disposed of at a commercial TSDF operation under a Resource Conservation and Recovery Act (RCRA) permit.

PROGRAMS FOR DISPOSING OF CAIS AND OTHER CHEMICAL WARFARE MATERIEL

Because of the large numbers of casualties from chemical weapons in World War I, the international community agreed to ban their use as part of the Geneva Protocol of 1925. Since then, however, chemical weapons have been used in war a number of times, by Japan (in China), by Italy (in Ethiopia), and by Iraq (against Iranian and Iraqi Kurdish citizens)

---

[5] In the context of this report, "neutralization" refers to a chemical reaction (hydrolysis) in which agent is converted into reaction products less toxic than the starting chemical.

[6] There is an apparent discrepancy between this list of anticipated CAIS recovery sites and Table 1-5, which classifies the former defense depot at Ogden, Utah, as a site for "no further action." Both the list and the table are from the report to Congress (U.S. Army, 1998a).

INTRODUCTION

TABLE 1-4 Recovered CAIS Currently in Storage

Location	Quantity and Type
Pine Bluff Arsenal, Arkansas	4,408 bottles, 891 mixed vials and bottles
Deseret Chemical Depot, Utah	575 bottles, 578 vials
Johnston Atoll Chemical Agent Disposal System, Johnston Atoll	59 PIGs
Camp Bullis, Texas	25 K955 bottles
Fort Richardson, Alaska	7 K941 PIGs
Redstone Arsenal, Alabama	1 K941 PIG, 1 K941 bottle

Source: Brankowitz, 1998.

(e.g., IOM, 1993). A religious cult in Japan has been accused of releasing the nerve agent GB in a Tokyo subway in 1995.

The recent Iraqi attacks spurred another attempt to abolish chemical weapons, the Convention on the Prohibition of the Development, Production, Stockpiling and Use of Chemical Weapons and on Their Destruction (commonly called the "Chemical Weapons Convention," or CWC). This treaty, together with a congressional mandate in 1986, has shifted attention in this country to the best way to destroy the inventory of chemical weapons in both the stockpile and non-stockpile categories.[7]

Chemical Stockpile Disposal Program

In November 1985, the U.S. Congress passed Public Law 99-145 requiring the destruction of all U.S. unitary chemical agents and munitions located at eight continental U.S. storage sites and on Johnston Island in the Pacific Ocean.[8] The materiel at these specific storage sites at the time the law was passed was defined as the chemical stockpile. In response, the Army, as the executive agent, established the Chemical Stockpile Disposal Program. The stockpile's unitary chemical agents, which are highly toxic or lethal, are stored either in bulk containers or in chemical munitions. The Army has already begun disposal operations at one continental storage site, the Tooele Chemical Agent Disposal Facility (TOCDF) in Utah, and at the Johnston Atoll Chemical Agent Disposal System (JACADS) on Johnston Island in the Pacific. The current status of these disposal operations is shown in Table 1-6. Another NRC committee, the Committee for the Review

---

[7] In addition to the Chemical Stockpile Disposal Program and the NSCMP, the Army has an Alternative Technology and Approaches (ATA) Program and an Assembled Chemical Weapons Assessment (ACWA) Program, which are investigating nonincineration-based technologies for the disposal of chemical materiel. The Army's Cooperative Threat Reduction Program helps foreign countries destroy their chemical weapons. The Chemical Stockpile Emergency Preparedness Program assists affected communities with emergency planning during disposal operations. Additional information on the Army's chemical materiel disposal programs is available on the Internet at <http://www-pmcd.apgea.army.mil/text/w_body.html>.

[8] The term "unitary" refers to a single chemical loaded in munitions or stored as a lethal material (NRC, 1996a, p. 6). A "binary" chemical weapon is "one that forms a lethal chemical agent from nonlethal constituents through a chemical reaction occurring during the time of flight to the target" (U.S. Army, 1996; p. 3-1).

TABLE 1-5 Potential CAIS Burial Sites, as Reported to Congress by the Army

State	Location	Site Classification[a]
Alabama	Redstone Arsenal	3, 4
Alaska	Unalaska Island (now closed)	3
	Gerstle River Test Site	2
Arizona	Yuma Proving Ground	1
Arkansas	Fort Chaffee	3
California	Fort Ord	4, 5
	Santa Rosa Army Airfield (now closed)	1
Georgia	Fort Benning	2
Hawaii	Schoefield Barracks	4
Indiana	Camp Atterbury	4
	Naval Surface Warfare Center, Crane Division	3
Iowa	Camp Dodge (now closed)	3
Louisiana	Camp Claiborne (now closed)	3
	England Air Force Base	2
	Fort Polk	3, 5
Maryland	Fort Meade	3
Massachusetts	Fort Devens (now closed)	1
Mississippi	Camp Van Dorn (now closed)	4
Missouri	Camp Crowder (now closed)	4
New York	Camp Hero (now closed)	4
	Plattsburg Air Force Base	1
North Carolina	Camp LeJeune Marine Corps Base	3
South Dakota	Ellsworth Air Force Base	3
Tennessee	Defense Depot Memphis	4
Texas	Camp Bullis	1
Utah	Defense Depot Ogden	4, 5

[a] 1 = known burial; 2 = likely burial; 3 = suspected burial; 4 = possible burial; 5 = no further action. Several other sites have been evaluated by the Army and classified as "no further action" required: Fort Wainwright, Alaska; Barksdale Air Force Base, Louisiana; Mitchell Field, New York (now closed); Defense Distribution Region East, Pennsylvania; Camp Barkeley, Texas (now closed); Fort Belvoir, Virginia; F.E. Warren Air Force Base, Wyoming. Camps and forts = Army bases.

Source: Adapted from U.S. Army, 1998a, Table D-1.

TABLE 1-6 Status of Agent Destruction at JACADS and TOCDF, as of April 25, 1999.

	JACADS[a]	TOCDF[b]	Total Stockpile[c]
Original tonnage	2,030 tons	13,616 tons	31,495 tons
Remaining tonnage	385 tons	10,939 tons	21,173 tons
Destroyed to date	1,645 tons (81.0%)	2,677 tons (19.7%)	4,322 tons (13.7%)

[a]Munitions and bulk containers destroyed: 13,020 GB nerve agent-filled projectiles (8 inch); 49,360 GB nerve agent-filled projectiles (105mm); 107,197 GB nerve agent-filled projectiles (155mm); 2,570 MK-94 (500 pound) bombs filled with GB nerve agent; 3,047 MC-1 (750 pound) bombs filled with GB nerve agent; 72,242 M-55 GB and VX nerve agent-filled rockets/warheads; 45,108 blister agent-filled projectiles (105mm); 68 blister agent-filled ton containers; 66 GB nerve agent-filled ton containers; 45,108 blister agent-filled projectiles (105 mm); 43,660 blister agent-filled mortars (4.2-inch).
[b]Munitions and bulk containers destroyed: 4,463 MC-1 (750 pound) bombs filled with GB nerve agent; 2,636 GB nerve agent-filled ton containers; 19,860 M-55 GB nerve agent-filled rockets; 116,045 GB nerve agent-filled projectiles (105 mm).
[c]Includes chemical warfare materiel at other stockpile sites (e.g., Aberdeen, Maryland; Newport, Indiana).

Source: U.S. Army, 1999.

and Evaluation of the Army Chemical Stockpile Disposal Program, has reported on and provided scientific and technical recommendations for various aspects of this program for more than 10 years (see, for example, NRC, 1994a, 1999a).

Non-Stockpile Chemical Materiel Program

Prior to 1991, the Army program for the disposal of chemical warfare agents was limited to the unitary chemical agents and munitions defined by statute as the "stockpile." The 1991 Defense Appropriations Act directed the Secretary of Defense to establish an office with the responsibility of destroying non-stockpile chemical materiel. The Project Manager for Non-Stockpile Chemical Materiel was assigned this task under the newly established U.S. Army Chemical Materiel Destruction Agency.

In the 1993 Defense Appropriations Act (Section 176 of Public Law 102-484), the Army was directed to report the locations, types, and quantities of non-stockpile chemical materiel; explain the methods to be used for its destruction; provide cost and time estimates; and assess transportation options. The *Survey and Analysis Report* provided an overview of the task facing the Army (U.S. Army, 1993, 1996). According to this report, non-stockpile chemical warfare materiel is located at more than 200 sites in the United States and in U.S. territories. At most sites, the materiel contains small quantities of chemical agent and does not appear to pose immediate hazards to the public or the environment. However, chemical weapons agreements and the continuing discovery of buried chemical warfare materiel have increased the impetus for locating and disposing of all non-stockpile chemical materiel.

The purpose of the NSCMP is to provide centralized management and direction for the destruction of non-stockpile chemical materiel, develop characterization and disposal facilities, develop schedule and cost estimates, and ensure compliance with federal, state, and local regulations. The NSCMP is responsible for the disposal or destruction of five types of chemical warfare materiel, each of which presents unique disposal problems: (1) buried chemical warfare materiel; (2) recovered chemical warfare materiel; (3) binary weapons; (4) former production facilities; and (5) miscellaneous chemical warfare materiel. Although CAIS are only relevant to the first two categories, all five are briefly described below to place the problem of CAIS disposal in the larger context of disposing of all types of non-stockpile materiel.

Buried Chemical Warfare Materiel

As of 1996, based on initial surveys, site visits, and interviews, the Army had located 168 potential burial sites at 63 locations in 31 states, the U.S. Virgin Islands, and the District of Columbia (U.S. Army, 1996). Of the 63 locations in the continental United States, most are current or former defense sites. In the interim version of the *Survey and Analysis* Report (dated April 1993), 224 potential burial sites were identified at 96 locations in 38 states (U.S. Army, 1993). A number of these sites have since been evaluated and characterized as requiring no further action.

Types of burial sites include (1) sites with CAIS only; (2) sites with small quantities of materiel (possibly including CAIS) with no associated explosives; (3) sites with small quantities of materiel with explosives; and (4) sites with large quantities of materiel with and without explosives. The majority of the sites have small quantities of materiel that may be treatable by transportable (mobile) disposal facilities. The treatment of explosively configured chemical warfare materiel will involve more hazardous operations. Larger quantities of materiel will probably be treated in fixed facilities. Large-quantity sites are located at four military installations: Aberdeen Proving Ground, Maryland; Deseret Chemical Depot, Utah; Rocky Mountain Arsenal, Colorado; and Redstone Arsenal, Alabama. Buried munitions include mortar rounds, bombs, rockets, projectiles, and bulk containers (55-gallon and "ton" containers). Chemical agents contained in these munitions include blister agents (sulfur mustard or lewisite), nerve agents (GA, GB, or VX), blood agents (hydrogen cyanide or cyanogen chloride), and a choking agent (phosgene). Other hazardous substances, such as white phosphorus, may also be present.

All sites will require one or more of the following steps: (1) site characterization via soil sampling, metal detection, the use of other ground-penetrating sensor technologies, and nonintrusive identification methods; (2) leaving the materiel in the ground, containing potential contamination, and controlling access to the site; (3) on-site treatment using transportable or fixed destruction facilities; or (4) transport of the materiel to another site for storage and destruction, if allowed by federal and state safety, transportation, and environmental regulations.

Recovered Chemical Warfare Materiel

This category of non-stockpile chemical warfare materiel includes munitions discovered during range-clearing operations, items previously removed from burial sites, and remnants from research and development activities. Much of this materiel is located at nine military installations: Aberdeen Proving Ground, Maryland; Dugway Proving Ground, Utah; Johnston Island, Pacific Ocean; Pine Bluff Arsenal, Arkansas; Redstone Arsenal, Alabama; Fort Richardson, Alaska; Fort Ord, California; Camp Bullis, Texas; and Deseret Chemical Depot, Utah. Recovered chemical munitions (mortar cartridges, artillery projectiles, bombs, and World War II German Traktor rockets) and containers of incapacitating agents and chemical agents (CAIS, unidentified glass bottles, and bulk containers) are included in this category. Many of the recovered items are the same or similar to those scheduled for destruction through the Chemical Stockpile Disposal

Program. However, some recovered items may be more difficult to destroy because of their deteriorated condition or the uncertainty of their contents.[9]

Compared with the amounts of chemical agent in buried munitions or bulk containers, the amounts in individual CAIS items (vials, ampoules, etc.) and, for most CAIS types, even in an entire set are small. A 155-mm projectile, for example, contains about 2.8 kg of agent, or about 2.3 liters if the agent is HD. The smallest HD-carrying projectile, a 105-mm M60 cartridge, contains 1.35 kg of HD, or about 1 liter. A ton container of HD holds 726 kg, or about 672 liters (U.S. Army, 1988). As Table 1-3 shows, only the two CAIS types that were used for decontamination training contained 2 to 3 liters of agent "as produced." The other CAIS types contained less than a liter as produced.

Following recovery from range-clearing operations or a burial site, recovered chemical warfare materiel is overpacked and either stored on site or transported and stored at a military site with an appropriate permit. After identifying the type and quantity of recovered materiel at a given site, the NSCMP conducts a destination analysis to support the decision to transport or store the materiel. If the decision is made to store it on site, the NSCMP prepares an Interim Holding Facility Plan. If the materiel is to be moved for storage and ultimate destruction, the Army prepares a transportation plan. The NSCMP considers risk to the public and the environment in deciding whether to store or transport the materiel. As required by federal law, the U.S. Department of Health and Human Services reviews the plans and recommends precautionary measures to protect public health and safety.

To handle various types of recovered non-stockpile materiel, the NSCMP is developing a number of disposal technologies. One is the RRS, described above and in Chapter 5, for disposal of CAIS sets or separated items. Three types of munitions management device (MMD) are planned for handling recovered munitions. In addition, the Emergency Destruction System (EDS) is being developed for materiel that is too dangerous to move. The EDS will use a shaped-charge explosive to open a munition casing or other container and then treat the agent by chemical neutralization. For details on the MMS and EDS, see Appendix C.

Binary Chemical Weapons

In binary chemical weapons, chemical agent is produced by a chemical reaction of two nonlethal components at the time the weapon is fired. Three types of binary weapons were tested: (1) the Navy Bigeye bomb; (2) a binary round for the Multiple Launch Rocket System, and (3) the M687 binary 155-millimeter GB projectile, which is the only binary chemical munition that entered full-scale production (production numbers are classified). The Bigeye bomb was not produced or stockpiled, but 200 test weapons and associated equipment are currently in storage at Pine Bluff Arsenal and must be destroyed. The Multiple Launch Rocket System binary munition development program was never completed (prototype development ended in October 1990 [U.S. Army, 1996]), but production equipment and completed components, located at Pine Bluff Arsenal, will require destruction.

The components of the M687 projectile are stored at Pine Bluff Arsenal; Umatilla Depot Activity, Oregon; and Deseret Chemical Depot, Utah (U.S. Army, 1993). The individual precursor chemicals that react to form the nerve agent GB are not classified as

---

[9]Recovered materiel is typically characterized by x-ray radiography (to look for explosives) and portable isotopic neutron spectrometry (to determine the presence and type of chemical agent); some materiel is extremely difficult to characterize because of its chemical condition.

chemical warfare agents, and they are stored separately. Therefore, they should present a relatively minor disposal problem. Some of the precursors (a liquid precursor called QL and powdered sulfur) that react in the Bigeye bomb to form the nerve agent VX must also be destroyed (U.S. Army, 1996). Simple disassembly operations, followed by chemical neutralization or thermal destruction of the precursors, are planned for the disposal of binary weapons.

Former Production Facilities

Seven former production facilities for chemical weapons were identified to be decommissioned: Pine Bluff Arsenal, Arkansas; Newport Army Ammunition Plant, Indiana; Aberdeen Proving Ground, Maryland; Rocky Mountain Arsenal, Colorado; Northrop Carolina Corporation Facility, Swannanoa, North Carolina; the Marquardt Facility, Van Nuys, California;[10] and the Phosphate Development Works, Muscle Shoals, Alabama.[11] These facilities produced BZ (an incapacitating agent), binary agents, unitary agents (VX, GB, mustard, and lewisite), or their precursors at various times from 1941 to 1990. Asbestos and polychlorinated biphenyls (PCBs) are also believed to be present at a number of these sites.

Destruction of these facilities is a three-phase process: (1) project definition, including contamination assessment; (2) prerequisite steps (e.g., abatement of asbestos and other hazards); and (3) demolition and disposal, which includes dismantling and destroying process equipment, plumbing, underground structures, and, if required, building materials, followed by environmental remediation of the surrounding site.

Miscellaneous Chemical Warfare Materiel

This category includes munitions, bulk containers, support equipment, and other devices that (1) were never filled with chemical agent, (2) were filled with simulants, or (3) were filled with agent but were later drained and refilled with decontaminating solution. The CWC requires the destruction of this materiel within five years after the agreement enters into force. (The CWC entered into force on 29 April 1997, 180 days after the deposition of the 65th instrument of ratification.) This materiel is currently stored at eight military facilities in the United States (U.S. Army, 1996).[12] Most of this materiel, which is not contaminated with agent, can be disposed of by traditional means. However, associated explosives, propellants, and agent simulants will require separate treatment.

LEGAL AND REGULATORY CONTEXT FOR CAIS DISPOSAL

The disposal of CAIS is constrained by a number of legal and regulatory issues. First, Army regulations currently classify CAIS as chemical warfare materiel, rather than as a

---

[10]Disposal plans for the Northrop Carolina Corporation Facility and the Marquardt facility, both commercial properties, are unknown.

[11]Disposal of this facility was conducted by the Tennessee Valley Authority beginning in the early 1980s. No significant structures remain.

[12]The interim version of the Survey and Analysis Report (U.S. Army, 1993) reported that miscellaneous chemical warfare materiel was located at Newport Army Ammunition Plant, Indiana, and Johnston Island in the Pacific Ocean.

characteristic hazardous waste under RCRA.[13] This classification mandates Army control of the materiel for all transportation and disposal, thus prohibiting the use of commercial facilities. However, whether CAIS or some of the chemicals contained in them are *lethal chemical agents* pursuant to the underlying statute, 50 U.S.C. 1512, is not clear.[14] Nonetheless, the conditions imposed by these regulations on the handling, transportation, and disposal of CAIS increase costs significantly but may not increase safety.

Second, some chemicals contained in CAIS (sulfur mustard and lewisite) are classified as chemical warfare agents, while others (e.g., phosgene and chloropicrin) are classified as industrial chemicals and hazardous waste, allowing them to be disposed of in commercial facilities. This distinction requires that recovered CAIS be unpacked and individual items characterized, segregated, and then repackaged for separate transport and disposal.

Third, under this interpretation of CAIS classification, the transport of CAIS requires government and regulatory approval. Prior to transport, the Army must prepare a transportation plan for approval of the Department of Health and Human Services and must obtain the approval of the governors of all states affected by the proposed transport, particularly the state in which the final destination is located.

Fourth, as noted above, the Army facilities built by the Chemical Stockpile Disposal Program for disposal of stockpile chemical materiel are prohibited by federal law from disposing of non-stockpile materiel (or any other hazardous waste except the declared stockpile). Although these facilities were specifically designed for the disposal of chemical agents, this prohibition prevents their use for CAIS disposal.

INTERNATIONAL APPROACHES TO CAIS DISPOSAL

The disposal of non-stockpile chemical materiel is an international problem. In Europe, farmers have been recovering WWI-era shells for many years, and fishermen have recovered chemical materiel from the North Sea and the Mediterranean. Despite the international dimensions of this problem, apparently no intergovernmental working groups are dealing with chemical weapons disposal or sharing "lessons learned." Active elements of the Technical Secretariat of the Organization for Prohibition of Chemical Weapons have focused on inspections rather than the development or sharing of technology. Army staff have attended a number of technical conferences sponsored by German, British, and NATO groups, but no intergovernmental working group has been established to address non-stockpile disposal issues. The continued involvement of the Army in international conferences and technical exchanges may provide a future forum for sharing information on destruction technologies, as well as on new and advanced detection and identification methodologies.

---

[13] Under RCRA, a substance is determined to be a hazardous waste either because it is listed as such in the law (a listed hazardous waste) or because its characteristics meet the conditions specified in the law for being treated as a hazardous waste (a characteristic hazardous waste). Based on the second criterion, sulfur mustard and lewisite are hazardous wastes.

[14] References to federal law in this report use the conventional format of the title of the United States Code (U.S.C.) followed by the section within that title. Appendix D discusses the law and the interpretation of it in Army regulations at greater length.

CAIS AND THE CHEMICAL WEAPONS CONVENTION

Chemical warfare materiel buried before 1972 is not covered under the CWC as long as it remains buried. However, recovered chemical warfare materiel, including recovered CAIS, normally has to be disposed of within 10 years. The United States, through a military working group, is pursuing a special classification for CAIS that would exempt them from the normal disposal schedule (Wakefield, 1999). (As of this writing, the working group has not released its report, so the rationale for an exemption is not known to the committee.)

Other nations have similar chemical agent training sets, and inspection protocols for their disposal have been a matter of discussion in CWC working groups. For example, prior to disposing of recovered CAIS items by incineration, the British treat them as recovered chemical warfare materiel subject to CWC requirements (Libby, 1999). The British use an incinerator, located on the military reservation at Porton Down, to dispose of all waste contaminated with or containing chemical agent, including CAIS. The Germans use a military incinerator at Munster to destroy CAIS recovered from German sites.

2

Disposal Alternatives

Chemical warfare materiel has been disposed of in various ways. Accepted practices once included open-pit burning, ocean dumping, and, most commonly in the case of CAIS, burial, either as-is or following field neutralization. Today these methods are not allowed, and other alternatives for CAIS disposal must be considered. (Recovered chemical warfare materiel that is too dangerous to transport because of its deteriorated condition can be disposed of on site. In these cases, personnel from the Army's Explosive Ordnance Detachment use explosives to destroy the materiel and consume the chemical agents.) The Army's baseline plan for the disposal of CAIS involves the RRS, a transportable disposal system that would be moved to CAIS recovery and storage sites. The primary treatment step would be chemical neutralization; in the present plan, this would be followed by commercial incineration of the neutralization wastes.

The Army is continuing the development of the RRS and has recently been permitted to begin testing operations. At the same time, the Army has explored the use of commercial facilities for CAIS disposal. Commercial facilities could be less expensive than the RRS, particularly for recovery sites with small quantities of CAIS. The Army documented its findings on using commercial facilities in its report to Congress (U.S. Army 1998a). The committee evaluated both the RRS and the use of commercial facilities for CAIS disposal, as well as other disposal alternatives.

ALTERNATIVES CONSIDERED

The following CAIS disposal alternatives are shown in Figure 2-1:

- Do nothing.
- Gather CAIS and store indefinitely.
- Use the baseline, mobile RRS for on-site treatment.
- Use the RRS in a fixed location.
- Develop a modified RRS (similar to the Expedient CAIS Disposal System [ECS] or EDS).
- Dispose of CAIS in commercial facilities.
- Dispose of CAIS in Army stockpile disposal facilities.

Do Nothing

One obvious disposal alternative is to do nothing (i.e., to leave the CAIS items buried and unrecovered). This alternative is usually one of the baseline options for the

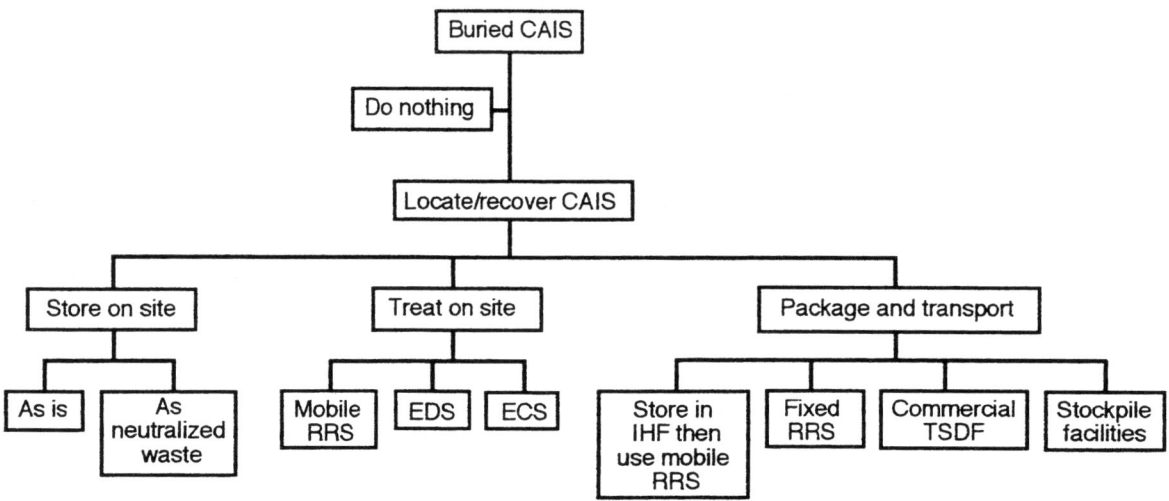

FIGURE 2-1 CAIS disposal alternatives. IHF = interim holding facility; RRS = Rapid Response System; ECS = Expedient CAIS Disposal System; TSDF = treatment, storage, and disposal facility; EDS = Emergency Destruction System.

cleanup of most hazardous waste. In some instances, for example when long-standing contamination will be remediated through natural attenuation or if the contaminants are immobilized in a controlled, secure, monitored, and properly permitted landfill, a "Do Nothing" approach may be a viable alternative to expensive remediation that would have limited long-term benefits. If the CAIS were known to be buried in controlled and monitored burial sites that would remain indefinitely under strictly enforced institutional controls, then a Do Nothing approach might incur less total risk (and especially less risk to soldiers and civilian workers) than recovery, treatment prior to transport, and final disposal. However, the Do Nothing option is not viable for CAIS disposal for the following reasons:

- Although the hazardous chemicals contained in CAIS may be remediated through natural attenuation when exposed to the environment, the chemicals are most likely to be found intact in their original glass vials or bottles, or even in their original shipping containers. These chemicals are likely to maintain their original chemical characteristics for the foreseeable future. (Although the plastic caps on some CAIS bottles have been known to degrade over time, allowing some exposure of their contents to the environment, some agents tend to form gels, which limit natural attenuation, when in contact with soil and/or subsurface moisture.)
- In many instances, CAIS are recovered on sites scheduled for release to civilian use or public access. Burial sites on property that is being developed by the public (e.g., former defense sites) could present health hazards to the local populace. These sites must be actively remediated to protect the community.
- Small accidental CAIS discoveries present a health risk to the untrained public and must continue to be actively treated as quickly as possible by Army Technical Escort Units.

Because the Do Nothing option is not a viable alternative for the known circumstances of CAIS discovery, the committee considered only CAIS *disposal* options.

Gather CAIS and Store Indefinitely

The Army could undertake an active program to locate and recover all known CAIS items and store them indefinitely at a permitted storage site, either as found or following preliminary treatment with currently available neutralization technology. While the CAIS are in storage, the Army could conduct research to develop new disposal technologies within the constraints of programmatic and CWC treaty deadlines. Upon discovery, the CAIS would be characterized, the chemical warfare materiel separated and repackaged, and all of the items sent to a permitted storage site for processing and destruction at some future time.

If no permit modifications are necessary to bring newly recovered CAIS into the states involved and if current storage permits apply, the permitting costs should be minor. Transportation costs of moving recovered CAIS would be about the same as the costs of moving the CAIS to a fixed RRS. Additional storage costs would be incurred, however. The cost of continued storage of 10 PIGs containing CAIS items is estimated by the Army to be $300 per day (at Fort Richardson, Alaska; Pine Bluff Arsenal, Arkansas; and Deseret Chemical Depot, Utah).[1] At that rate, storage costs would be more than $100,000 per year. The cost of identifying and characterizing CAIS materials, separating the industrial chemicals from the chemical warfare agents, and repackaging would be the same as for the RRS alternatives. Issues of processing cost and cost recovery would not apply.

Baseline Rapid Response System

The Army's baseline approach is the RRS for on-site treatment of CAIS items. The RRS is a mobile unit designed specifically to dispose of CAIS items at the locations where they are found.[2] The RRS operations unit contains a series of linked glove boxes[3] equipped to remove CAIS ampoules and bottles from their packages, identify their contents, and then segregate and repackage CAIS containing industrial chemicals for off-site commercial disposal. Only CAIS containing sulfur mustard or lewisite would be treated in the RRS. Within the glove boxes, the glass containers are crushed in a reactor containing a chemical that rapidly neutralizes the chemical agent. The contents of the reactor (reagent, solvents, agent degradation products, and glass fragments) are then transferred to a sealed container for treatment at a commercial TSDF before final disposal.

The RRS has recently been permitted by the state of Utah to begin an initial test program with both simulants and chemical agents at the Deseret Chemical Depot. Once the RRS has been successfully tested, operational deployments can begin. One site considered for an early operational deployment of the RRS is Fort Richardson, Alaska,

---

[1] The basis for this estimate is not detailed in the report (see U.S. Army, 1997a).

[2] For additional details, see Appendix C and the Internet web site for the NSCMP: <http://www-pmcd.apgea.army.mil/text/NSCMP/IP/FS/RRS/ index.html>.

[3] A "glove box" is a sealable container with transparent sides or observation ports. It has two or more access ports to which long rubber or plastic gloves are attached for manipulating items inside the sealed space.

TABLE 2-1 Commercial Incinerator Facilities with Hazardous Waste Permits[a]

Operating Facilities	Facilities Permitted but Not in Operation	Facilities with Permit Applications Pending
Safety Kleen (Laidlaw Environmental Services), Bridgeport, New Jersey	Giant Cement, Harleyville, South Carolina	Organic Incineration Technologies, Fairbanks, Alaska
Safety-Kleen, (Laidlaw Environmental Services), Clarence, New York[b]	Reynolds Aluminum, Gum Springs, Arkansas[c]	Grant County Waste Management, Beverly, Washington
Allied Chemical, Birmingham, Alabama[d]	American Envirotech, Channelview, Texas	Scientific Ecology Group, Oak Ridge, Tennessee
Safety-Kleen (Laidlaw Environmental Services), Roebuck, South Carolina (scheduled for closure)	Houston Chemical Services, La Porte, Texas	
Thermal KEM, Rock Hill, South Carolina (scheduled for closure)	GTX, Morgan City, Louisiana	
LWD, Inc., Calvert City, Kentucky		
Atochem, Carrollton, Kentucky[e]		
Chemical Waste Management (TWI), Sauget, Illinois		
Chemical Waste Management Chemical Services, Chicago, Illinois (scheduled for closure)		
Waste Research & Reclamation, Eau Claire, Wisconsin		
Ross Incineration Services, Inc., Grafton, Ohio		

TABLE 2-1 Continued

Operating Facilities	Facilities Permitted but Not in Operation	Facilities with Permit Applications Pending
Waste Technologies Industries (WTI), East Liverpool, Ohio		
Rhone-Poulenc Basic Chemical Co., Baton Rouge, Louisiana		
Safety-Kleen, (Laidlaw Environmental Services), Coffeyville, Kansas[f]		
Safety-Kleen, (Laidlaw Environmental Services), Aragonite, Utah		
ICI, Joplin, Missouri		
Huges Environmental, Brookville, Mississippi		
Clean Harbors, Kimball, Nebraska		
Safety-Kleen, (Laidlaw Environmental Services), Deer Park, Texas		
Chemical Waste Management Port Arthur, Texas		
Safety-Kleen, (Laidlaw Environmental Services), Clive, Utah (scheduled for closure)		

[a]These commercially operated incineration facilities have permits to receive hazardous wastes (as defined by EPA) from other sites. Each facility has site-specific requirements imposed by EPA or a state regulator. Neither EPA nor the committee has evaluated whether any of the existing permits would allow or prohibit disposal of CAIS at that facility. The committee received information from an NRC report reviewer that many additional operating cement kilns have permits to incinerate hazardous waste.
[b]Limited to ignitable, corrosive, and reactive wastes.
[c]Limited to F088 waste streams.
[d]Limited to wood-preserving and coal-tar wastes.
[e]Limited to high-BTU and high-tin waste streams.
[f]Pending permit application for new unit.

Source: Updated from EPA, 1999, with information from an NRC report reviewer.

where seven PIGs containing CAIS have been recovered. The Army compared the estimated risks and costs of RRS deployment to Fort Richardson with the risks and costs of transporting the CAIS items from Alaska to Utah for disposal in the RRS (U.S. Army, 1997a). The committee discusses this cost analysis in Chapter 4.

Fixed-Mode Rapid Response System

The committee also considered the use of an RRS in a fixed mode at one or more regional sites. Recovered CAIS items would be sent to the RRS(s) for disposal at these sites. The operation of the RRS itself would be identical to the mobile RRS option, except that the startup and shutdown phases would be simplified because the unit would remain in place. The transportation phase would be eliminated. Because the CAIS rather than the RRS and its associated equipment would be transported, there could be differences from the mobile RRS in costs, permitting requirements, risks, and public reaction.

Modified Rapid Response System

Modified RRS equipment could also be developed. One example is the ECS, which the Army is considering (U.S. Army, 1998b). The ECS is essentially a mobile glove box that can treat loose CAIS vials not found in PIGS. Once CAIS are recovered, regulatory requirements mandate that they be disposed of in less than 90 days or that a storage permit be obtained. The ECS would be deployed to a recovery location where CAIS items would be disposed of in less than 90 days. Use of the ECS is intended to eliminate the need for a storage permit.

A mobile glove box such as the ECS would not have the characterization capability of the RRS. Therefore, CAIS recovery personnel (Technical Escort Units and the U.S. Army Corps of Engineers) would have to be equipped with portable Raman spectroscopy to identify and separate chemical warfare materiel from industrial chemicals in unearthed CAIS vials and portable isotopic neutron spectroscopy for identifying the contents of PIGS (see Appendix C for details).

Commercial Disposal

The Army outlined an approach for disposing of CAIS at commercial facilities in its report to Congress and in a supplementary technical report (U.S. Army, 1998a; Amr et al., 1998). Excerpts of the Army report are provided in Chapter 4, which focuses on the commercial disposal as presented in the Army report.

A list of commercial incineration facilities for the disposal of hazardous waste is given in Table 2-1. Additional facilities at various industrial and government sites only treat on-site waste materials. As explained in the discussion of the study task (see Chapter 1), the committee discovered during its investigation of the commercial disposal option for CAIS that commercial facilities would probably use incineration-based disposal methods. Therefore, the committee focused its evaluation of the commerical disposal option on incineration.

Commercial disposal could involve either incineration-based or nonincineration-based disposal technology. Nonincineration methods include neutralization, biodegradation, wet-air oxidation, supercritical-water oxidation (SCWO), and possibly plasma arc or other thermal methods. Incinerators are defined in 40 CFR 260.10 as "any enclosed device that: (1) uses controlled flame combustion and neither meets the criteria for classification as a boiler, sludge dryer, or carbon regeneration unit, nor is listed as an industrial furnace; or (2) meets the definition of infrared incinerator or plasma arc incinerator." In an evaluation of nonincineration alternatives, it is important to consider the complete disposal system because in some cases an initial nonincineration step in the process is followed by an incineration step.

Nonincineration-Based Stockpile Disposal Facilities

The facilities being developed for the Chemical Stockpile Disposal Program, which are specifically designed to destroy chemical agent materials (see Chapter 1), offer some potential technological and economic benefits for CAIS disposal. However, many public and stakeholder groups are opposed to using stockpile facilities for other purposes, and current legal restrictions prohibit using them for the disposal of any other wastes, including CAIS or other non-stockpile chemical materiel. Furthermore, Army officials have publicly assured local residents living near stockpile disposal facilities that no other material will be disposed of at these locations. Thus, there are significant nontechnical prohibitions, which would have to be appropriately addressed through a public involvement program and congressional action, against the use of stockpile facilities for CAIS disposal.

The stockpile facilities of interest to the committee are those that use nonincineration technology for destruction of sulfur mustard or lewisite. The Army already has a pilot-plant facility, the Chemical Agent Munitions Disposal System in Utah, equipped to dispose of lewisite in the small quantities found in CAIS. This facility and the neutralization facility planned for Aberdeen Proving Ground, Maryland, also have or will have technology to destroy sulfur mustard. A significant advantage of these sites is that their use would ensure the use of an acceptable method with good process controls and appropriate safety precautions (monitoring of workers and effluent streams, safe reception and unpacking operations, etc.). Nearby communities and other interested groups may be amenable to the use of these facilities for destroying CAIS found at that site or even CAIS found elsewhere in the same state. Transporting significant numbers of CAIS sets or items from out-of-state sites seems more problematic.

ALTERNATIVES SELECTED FOR ANALYSIS

After reviewing all of these alternatives, the committee chose to focus on the following technical options: (1) commercial disposal by incineration; (2) baseline, mobile RRS; (3) fixed RRS; and (4) nonincineration-based methods.

3

Issues to Consider

The committee developed a list of issues to be considered in evaluating CAIS disposal alternatives. The issues were organized under six headings: technology; laws and regulations; costs; environmental impacts, worker/public safety, and risks; public involvement; and programmatic aspects.

TECHNOLOGY

The major technical challenge is to establish that a disposal process can destroy chemical agents safely, reliably, and effectively and that the destruction products can be disposed of safely, reliably, and effectively.

Process Reliability and Effectiveness

Any CAIS disposal method must demonstrate the capability of (1) destroying chemical agents to treaty and regulatory standards; (2) operating safely and effectively with varying agent feed rates and under nonroutine conditions, such as might occur during a power failure or severe weather; (3) handling a variety of chemical agents and containers, some of which may be badly deteriorated; and (4) decontaminating or destroying the agent containers.

Technical Maturity of the Process

One way to minimize technical risks and development costs is to use a process that has already been used successfully and extensively. Established processes, such as the baseline incineration process for stockpile destruction or the nonincineration processes in use commercially for some industrial wastes, entail much less technical risk than processes that are still under development.

Monitoring and Disposal of Process Effluents

Besides the selection of an effective and reliable process, safe CAIS disposal will require monitoring and disposal of process effluents. All disposal technologies produce gaseous, liquid, and solid effluents in varying proportions. Ideally, each effluent can be contained until it has been analyzed and certified safe for release into the environment. In practice, given the small scale of the CAIS disposal operations, some CAIS-derived effluents may be produced in quantities too small to be monitored effectively or efficiently in the total effluent stream(s) from a waste disposal process, but also too small to present significant hazards. For example, when an HD-containing CAIS sample is burned in a commercial incinerator, much supplemental fuel will be needed to maintain an adequate combustion temperature. The amount of HD combustion products in the stack gas will therefore be negligible.

LAWS AND REGULATIONS

The legal and regulatory context for the CAIS disposal problem was described in Chapter 1. In evaluating a particular disposal alternative, the mutual consistency of the existing laws, regulations, and treaties must be considered (see Box 3-1). One issue is the current classification of CAIS sets and items as chemical warfare materiel and whether they could be reclassified as a characteristic hazardous waste under RCRA (the Resource Conservation and Recovery Act). Special requirements for transporting chemical warfare materiel apply to CAIS sets and items under the current classification. A second issue is that two CAIS chemicals, sulfur mustard and lewisite, are classified as chemical warfare agents, while other CAIS chemicals are classified as industrial chemicals and hazardous waste. A third issue for CAIS disposal is that Army facilities built for the Chemical Stockpile Disposal Program are prohibited by federal law from being used to dispose of any materiel in the "non-stockpile" category, including CAIS sets, items, or chemicals.

The committee believes that disposal options that require extraordinary legal or regulatory changes will encounter significant hurdles. However, the key to resolving these issues with a consistent approach that protects workers, the public, and the environment is to classify complete CAIS or items separated from CAIS as a characteristic hazardous waste under RCRA, even if some of the chemicals found in CAIS, such as HD, continue to be classified as chemical warfare agents. This approach would be consistent with historical practice in environmental regulation. For example, many wastes are classified as solid wastes, not hazardous wastes, although they contain the same chemicals as hazardous wastes. The relative amount of the hazardous constituents and the risk associated with them are the basis for the difference in classification. The same substances present at higher levels would require that the waste be classified in the more stringent category of hazardous waste.

A reclassification of CAIS also makes sense from the perspective of the history of CAIS production and use. CAIS were intended to be used not as chemical weapons but as test kits for training troops to defend themselves from chemical attack. Thus, it is reasonable to regulate CAIS on the basis of the risks they pose, rather than as former chemical weapons. Federal laws and international treaties governing treatment of munitions and chemical weapons are not clear about whether CAIS must be categorized as chemical weapons and chemical warfare materiel (see Appendix D). The Army has very strictly construed the statutory scheme and classified CAIS as a lethal chemical warfare agent or chemical warfare materiel in its regulations and guidance documents. This classification brings with it prohibitions, constraints, and administrative requirements that greatly increase the cost of destroying CAIS but provide a negligible increase in safety to workers, the public, or the environment.

The EPA reviewed the chemical agents found in CAIS and concluded that they have the characteristics of hazardous wastes as defined by RCRA. EPA considers the federal

> BOX 3-1 Case Study: CAIS Recovery at the Raritan Arsenal
>
> A non-stockpile remediation action was conducted at the former Raritan Arsenal, New Jersey, from October 1995 through May 1996. Demilitarized and leaking chemical unexploded ordnance (UXO) and CAIS vials were discovered commingled at this site (DiMichele, 1999). The deteriorated and leaking UXO had contaminated more than 12 tons of soil with neat sulfur mustard and lewisite. The contaminated soil was treated on site by mixing it in a concrete mixer with a 10 percent calcium hypochlorite decontamination slurry solution for a minimum of 15 minutes. The treated soil was then packaged, shipped to a commercial disposal facility, and incinerated as hazardous waste.
>
> Because of regulatory requirements, the intact CAIS vials containing sulfur mustard or lewisite were handled very differently from the soil. Because these vials are categorized as chemical warfare materiel, they were packaged in a protective overpack by personnel from the Army's Technical Escort Unit, temporarily stored in an interim holding facility, and then shipped to the Army's facility at Pine Bluff, Arkansas, where they are currently being stored, while awaiting disposal by the RRS or an alternate disposal method.
>
> This case study demonstrates the inconsistency of the current regulatory requirements. Pure mustard agent that has leaked into the surrounding soil, which is extremely hazardous, can be treated on site and sent to a commercial hazardous waste incinerator for final disposal. CAIS in intact vials, which are by comparison easy to overpack and ship and were originally developed for use in training exercises, are subject to more stringent requirements. Had the CAIS vials been broken or leaking, the remains of the vials and the contaminated soil could have been shipped and disposed of in a commercial incinerator as hazardous waste contaminated media. If existing regulations were changed, intact CAIS sets or items could be handled in the same manner as leaking CAIS items and contaminated soil.

hazardous waste disposal requirements to be appropriate for handling CAIS chemicals safely. Furthermore, the permitting process under RCRA allows location-specific and chemical-specific conditions to be developed and made legally binding. Therefore, although CAIS may be stored, disposed of, or treated at a federally permitted hazardous waste TSDF, additional permitting requirements may also be imposed because of the specific characteristics of this waste.

In summary, the statutory and regulatory requirements under which CAIS are now treated as chemical warfare materiel were designed for munitions configured with agent and explosives or for large quantities of chemicals in bulk containers. The federally mandated system for cradle-to-grave handling, transport, and disposal of hazardous wastes already applies to CAIS chemicals and, in the committee's view, is a far more efficient and effective framework for CAIS disposal that would provide the same level of protection to workers, the public, and the environment. Shifting the framework controlling CAIS disposal from that of munitions and chemical weapons to one of characteristic hazardous waste under RCRA would therefore be reasonable and desirable. This shift may be feasible through a review and change in the Army's interpretation of the statutory language, or it may require clarification or amendment of that language by Congress.

Besides legal and regulatory requirements, the two principal components of CAIS, sulfur mustard and lewisite, which are currently classified as chemical warfare agents, are subject to administrative controls called "surety." Surety procedures ensure the safety, security, reliability, integrity, and authentication of complex, high-consequence systems, such as chemical or nuclear weapons. The position taken by both the NSCMP and the U.S. Army Corps of Engineers is that recovered CAIS are not subject to surety requirements. However, the current small-quantity exemption arguably does not apply to the quantities in some CAIS items, so formal clarification by the Army, perhaps through a specific exemption for recovered CAIS, would be helpful.[1]

COSTS

CAIS disposal costs are driven by five factors: the nature of recovered CAIS (sets and/or items); who disposes of them; and where, how, and when the disposal occurs. The cost of disposing of recovered loose bottles and vials will differ from the cost of disposing of bottles and vials inside PIGs or other metallic overpacks, which will require cutting. The waste streams and disposal procedures involved will also differ and will affect costs.

CAIS can be disposed of by a contractor in a commercial facility, by a contractor in a government-owned transportable facility (e.g., the RRS), or by Army personnel in a government-owned and operated facility. Cost estimates for commercial and government-owned equipment have differed by as much as two orders of magnitude. Therefore, the identity of the party disposing of CAIS can have a significant impact on cost.

If CAIS are disposed of at the location where they are recovered, costs will be incurred in bringing a disposal facility to the site and storing the items in an interim holding facility. If the items are disposed of at a permitted storage site, such as at Pine Bluff Arsenal in Arkansas, costs will be incurred in moving the CAIS items to this facility. If the CAIS are disposed of at a fixed commercial facility located away from the storage site, costs will be incurred in moving the CAIS items to the commercial facility, either from temporary storage at the recovery location or from permanent storage at a permitted site.

Treatment processes include incineration, neutralization followed by either incineration or another treatment of process wastes, or alternative processes, such as thermal reduction or SCWO (supercritical-water oxidation). Extended periods of storage and monitoring would be required for processes that are not currently in commercial operation.

Storage following the recovery of CAIS items is allowed for only 90 days without a RCRA permit (see Appendix D). After that, the items must be moved to a permitted storage facility, imposing additional costs for transport, storage, and monitoring. Also, if on-site treatment of the CAIS items is not completed within 90 days, a RCRA permit for CAIS disposal is required. Obtaining a RCRA permit has substantial cost implications.

Within these broad parameters, a large number of storage, disposal, and transport options are available. The major costs are permitting, facility modifications, transportation, processing operations, and indirect costs.

---

[1] Army Regulation 50-6 (U.S. Army, 1995b) includes surety exemptions for small quantities of research chemical agents, including quantities of pure HD and lewisite—25.0 ml each—and dilute solutions of HD and lewisite—100 and 50 mg, respectively. These exemptions would be exceeded by a single recovered CAIS bottle of the Toxic Gas Set, which contains 103.3 or 112.5 ml of neat sulfur mustard, or by individual, dilute CAIS vials/ampoules (5 percent solutions), which contain at least 2 ml of sulfur mustard or lewisite. Typical CAIS recovery sites, which involve multiple CAIS items, would far exceed these exemptions. Reportedly, some individuals have recognized the apparent inconsistency in the surety requirements between as-produced CAIS and buried/recovered CAIS, which may be as potent as the original CAIS chemicals.

Permitting

The costs of obtaining a RCRA permit, permits for discharging process effluents, and other permits and of complying with these permits can be considerable. Once a permit or permit modification is obtained, the costs imposed by the permit conditions can be considerable regardless of whether the CAIS are disposed of in a government or a commercial facility. Permit conditions that can affect processing costs include: (1) limitations on the processing rate for CAIS items; (2) a requirement that CAIS items be processed separately from other hazardous wastes; (3) requirements for monitoring emissions and monitoring agent; (4) requirements for reporting, which may include training plans, shutdown plans, contingency plans, quality control plans, and other documents; (5) staffing requirements, which may specify the number, skills, and training of personnel who handle CAIS items; and (6) requirements for waste disposal.

For a commercial facility, the cost of obtaining a permit modification to process chemical warfare materials, such as sulfur mustard or lewisite, and the attendant publicity and impact on community relations may be strong disincentives to accepting CAIS materials.

Facility Modifications

The cost of modifications for commercial facilities could adversely affect a company's willingness to accept CAIS items. Modifications could include: the addition of an area for receiving and unpacking CAIS items; the installation, testing, maintenance, operation, and calibration of agent monitors; and the addition of an area for packaging CAIS wastes. These costs could be amortized for large commercial facilities. A dedicated Army facility, such as the RRS, will not require modifications.

Transportation

If CAIS were not classified as lethal chemical warfare agents, they could be transported by commercial firms in accordance with federal hazardous materials transportation regulations (49 CFR, Parts 100-185). The cost of transport, as well as of segregating, characterizing, and repackaging recovered CAIS items are typically borne by the commercial hazardous waste disposal firm. In addition to the federal regulations for the transport of hazardous materials, state permitting authorities may impose additional requirements, which could increase costs. As long as CAIS are classified as lethal chemical warfare agents, established transport requirements apply and a transportation plan must be developed and approved by the Department of Health and Human Services.

If a transportable disposal facility, such as the RRS, is brought to the CAIS items, arrangements for transporting the facility must be made, and the logistics and costs of moving the equipment must be considered. For example, the RRS can either be driven or flown to locations where CAIS have been recovered. Transporting the RRS by air would require two C-141 transport aircraft, one for the RRS operations and utility trailers and one for transporters, a supply trailer, and a mobile analytical support laboratory. These transportation costs, as well as the cost of personnel, could be substantial and could have an impact on the Army's disposal decisions. If a commercial firm processes the CAIS items in a fixed facility, such as an incinerator, the costs of transportation to the facility are typically borne by the firm and would therefore be built into the fee charged the Army.

Processing Operations

Once the necessary permits and arrangements have been made for CAIS disposal, substantial preprocessing, processing, and postprocessing costs could be incurred by both private entities and the Army. These costs could reduce the savings the Army expects to realize by using commercial processing facilities. The cost elements for disposal in an Army-owned, contractor-operated facility include: (1) mobilization and site preparation; (2) direct labor and overhead for disposal operations; (3) demobilization and site shutdown; (4) disposal of CAIS waste streams; and (5) utilities, materials, and supplies.

For CAIS disposal using commercial facilities, operating costs depend on the quantities of CAIS materials received, the costs of waste characterization, the operator's ability to process CAIS items with other hazardous wastes rather than separately, facility preparation costs, specialized handling requirements, taxes and fees, waste stream disposal, facility decontamination (if required), and direct labor and overhead associated with CAIS disposal.

Indirect Costs

Indirect costs include engineering, administrative, and management support, the recovery of the costs of design and construction of the equipment, maintenance support, laboratory support, increased liability, and other items not directly related to the number of CAIS items processed. Indirect costs are factored into the costs of processing CAIS items in government-owned facilities but may not be uniformly included in estimates for commercial options.

ENVIRONMENTAL IMPACTS, WORKER/ PUBLIC SAFETY, AND RISKS

Environmental Impacts

The Army is in the process of preparing a programmatic environmental impact statement to address the potential effects of disposal operations, including potential accidents, on the environment, the ecosystem, and human health. The environmental effects on soil, groundwater, and air can be estimated from the properties of sulfur mustard and lewisite (see Table 3-1), the primary CAIS materials of concern in this study.

The physical state and water solubility of chemicals are factors that affect their environmental impacts. The melting point of pure sulfur mustard (bis[2-chloroethyl] sulfide) is 14.4°C. Sulfur mustard has a low vapor pressure, even at room temperature, and evaporates very slowly in cold climates. It also has low solubility in water. Shaw and Cullinane (1998) have shown that sulfur mustard is absorbed into surface materials. The combination of low solubility and a tendency to be absorbed suggests that it is not very mobile in a water environment (e.g., groundwater system). Neither Great Britain nor Canada has detected sulfur mustard in groundwater under their firing ranges (Shaw and Cullinane, 1998).

Sulfur mustard is also subject to hydrolysis (Shaw and Cullinane 1998). The EPA has determined that sulfur mustard reacts rapidly with water to form hydrogen sulfide and other compounds with significant toxicity, although they are less toxic than sulfur

TABLE 3-1 Properties of Sulfur Mustard and Lewisite

Chemical	Boiling Point	Vapor Pressure (mm Hg @ 20°C)	Vapor Density	Solubility in Water	Specific Gravity (@ 20°C)	Freezing Point	Volatility (mg/m^3 @ 20°C)
Lewisite	190°C	0.394	7.1	insoluble	1.89	0.1–18°C (depending on purity)	4,480
Sulfur Mustard	217°C	0.072	5.5	negligible	1.27	14.45°C	610

mustard (Amr et al., 1998). However, dissolved sulfur mustard reacts via hydrolysis, which means that the decomposition proceeds very slowly because of its low solubility. Biodegradation of hydrolyzed sulfur mustard has been carried out during agent decontamination (Shaw and Cullinane, 1998; NRC, 1996a), which suggests that biodegradation in the environment (natural attenuation) may be possible.

The solubility of pure lewisite (dichloro [2-chlorovinyl] arsine) in water is approximately the same as for sulfur mustard, but the volatility is higher. Hydrolysis in water is faster than for sulfur mustard. Although the low solubility of lewisite suggests that it would not be mobile in a water environment, its arsenic-containing hydrolysis products could present serious environmental problems if they are not immobilized by chemical or physical treatments.

Under the Clean Air Act, the air emissions from the disposal of CAIS material would require a permit. The permit would specify allowable emission rates, design criteria, operating criteria, monitoring requirements, and other measures to ensure compliance with permit requirements.

Worker/Public Safety

Various disposal alternatives raise different safety concerns for workers and the general public. Safety issues can be related to (1) handling, identification, and repackaging operations; (2) storage and monitoring operations; (3) transportation operations; and (4) unpackaging, treatment, and waste handling operations. These safety issues can be assessed on the basis of existing workplace safety standards (see Box 3-2).

The two principal CAIS chemicals of concern, sulfur mustard and lewisite, can be lethal following inhalation. Both can also be absorbed through the skin and can cause systemic effects, such as pulmonary edema, diarrhea, weakness, hypotension, and death. The effects of lewisite are more rapid in onset than those of sulfur mustard, but sulfur mustard is lethal at lower concentrations. Both are also carcinogens.
Many hazardous and toxic substances are routinely disposed of by commercial incineration or other means. In Figure 3-1, the acute lethal concentrations of nitrogen mustard and other highly hazardous industrial compounds are compared with those of the chemical warfare agents GB, sulfur mustard, lewisite, and other CAIS components (cyanogen chloride and chloropicrin). The highly hazardous industrial chemicals were selected from compounds listed in the *North American Emergency Response Guidebook*

ISSUES TO CONSIDER 45

> BOX 3-2 Workplace Exposure Standards
>
> Neither the Occupational Safety and Health Administration nor the American Conference of Governmental Industrial Hygienists has established workplace exposure standards for chemical warfare agents. However, workplace exposure limits developed by the Army have been independently reviewed and endorsed by the Centers for Disease Control and Prevention (CDC, 1988). The workplace standards for chemical warfare agents and the military regulations for applying these standards are documented in Army Regulation 385-61 (U.S. Army, 1997d).
>
> The current eight-hour time-weighted average (TWA) airborne exposure limit (AEL) is 3 $\mu g/m^3$ for sulfur mustard (HD) and lewisite. Although the AELs have been developed as permissible eight-hour TWAs for unmasked workers, the maximum concentrations to which workers may actually be exposed is limited to 20 percent of the AEL (0.6 $\mu g/m^3$). In practice, a control room pre-alarm sounds if airborne levels reach this level. Below this concentration, workers are not required to wear masks. At airborne concentrations higher than 0.2 times the AEL, workers must put on protective equipment and respirators. At 0.7 to 0.8 times the AEL, a local alarm sounds and workers must evacuate the area. (Only emergency response personnel can be present, and they must wear a supplied-air respirator or self-contained breathing apparatus [SCBA]). Although Army Regulation 385-61 requires the use of SCBAs for workers at HD or lewisite concentrations higher than the AEL, in practice, they are not allowed to remain in areas where airborne concentrations exceed the AEL.
>
> Military experience indicates that the standards, when applied as described above, have protected workers. However, there are no data demonstrating that this would be the case if the AELs were treated as true TWAs (that is, as average air concentrations to which workers could be exposed continuously without protective garb or respirators). Thus, if commercial facilities were used for CAIS disposal, the exposure levels for donning protective clothing and gear and for evacuating contaminated areas should be equivalent to the Army's practices and standards to ensure the same level of worker safety.

as requiring initial isolation distances of at least 400 feet from the source of small spills (DOT, 1996). Chlorine, a common industrial chemical, is included to provide a frame of reference.

Only two of the chemicals listed, GB and hydrogen selenide, are more lethal than sulfur mustard. Lewisite is the sixth most toxic compound. Thus, the lethalities of the chemical warfare agents present in CAIS are equivalent to or greater than the lethalities of these highly hazardous industrial chemicals. (Of course, the sulfur mustard and lewisite in many types of CAIS are in dilute forms.) The lethal inhalation doses of nitrogen mustard and sulfur mustard are similar. However, sulfur mustard is more toxic than nitrogen mustard by skin exposure to vapor. These similarities in lethal concentrations, as well as in vapor concentrations that cause severe effects following inhalation (data not shown), imply that facilities licensed to dispose of nitrogen mustard are probably adequately equipped to dispose of sulfur mustard.

If CAIS (or waste by-products associated with chemical neutralization of CAIS material) are reclassified as hazardous materials,[2] transportation requirements would be

---

[2] For a description of the debate over whether CAIS fall under the definition of "lethal chemical warfare agents," see Schmauder, 1997.

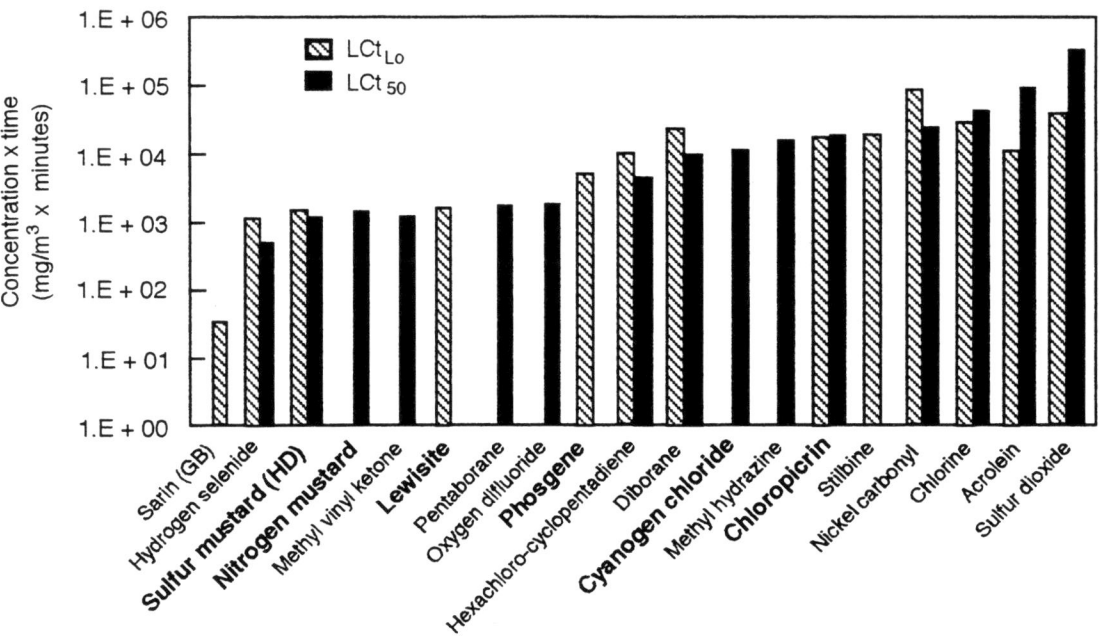

FIGURE 3-1 Comparison of acute lethal concentrations of CAIS chemicals and some highly toxic industrial chemicals. CAIS chemicals are shown in boldface (see Tables 1-1 to 1-3 in Chapter 1 for further information). LCt_{Lo} is the lowest tested concentration (mg/m^3) that caused fatalities, multiplied by the duration of exposure in minutes. LCt_{50} is the concentration (mg/m^3) that killed half of the test animals, multiplied by the duration of exposure in minutes.

Sources: The lethal concentration for GB is from NRC, 1997. The data for cyanogen chloride and nitrogen mustard are from U.S. Army Material Safety Data Sheets. All other data are from the RTECS database of the National Institute of Occupational Safety and Health.

set by the Hazardous Materials Regulations (49 CFR, Parts 100-185) and would not require that CAIS be transported under military controls. Consideration of technical feasibility could be incorporated in the transportation requirements through the use of established regulatory concepts such as ALARA (as low as reasonably achievable) or ALARP (as low as reasonably practicable). If CAIS components continue to be classified as lethal chemical warfare agents, they will be subject to 50 USC 1512, in which case, the Army has defined the transportation requirements (Fatz, 1997):

> Recovered non-stockpile CWM [chemical warfare materiel] will be overpacked in Army-approved containers that meets Department of Transportation (DOT) packaging regulations (49 CFR parts 172–178) and will be properly placarded, labeled, and manifested prior to transport off site. Off-site transport of recovered non-stockpile CWM shall be in accordance with 50 USC 1512-517. Off-site transport requires a transportation plan developed by the Program Manager for Chemical Demilitarization (PMCD), in coordination with the Commander of CBDCOM, and approved by the U.S. Department of Health and Human Services (USDHHS). Emergency removal activities may be initiated before, but off-site transportation of recovered non-stockpile CWM is not permitted until this plan has written approval by USDHHS. Audit trails of all non-stockpile CWM transportation and receipts of such non-stockpile CWM shall be established.

Packaging and shipping are further discussed in the 14 April 1998 U.S. Department of Transportation Approval CA-9804018, which requires that recovered chemical warfare materiel be transported under military control (e.g., by U.S. Army Technical Escort Unit personnel).

Risk Analysis

Each option for CAIS disposal, including the Do Nothing or Store Indefinitely options, poses some degree of risk to the environment, the public, and workers. The activities involved in the recovery, transportation, and disposal/ treatment processes pose different types and levels of risk. A variety of design features, procedures, training, and other measures can be put into place to reduce or control the risks of a particular option. The Army must consider the nature and levels of risk and the degree to which it can be controlled or reduced before selecting an option. For example, cost estimates can take design changes and control features into account.

The Army's consideration of risks associated with CAIS disposal options should be comprehensive (covering all activities and types of risk), comparable (treating each option equitably), and meaningful (focusing on significant factors based on the available data). Whether the analysis is qualitative, quantitative, or a combination of the two, each option should be reviewed based on identifying and understanding the risks of each process step, examining possible risk control measures, and putting these risks into context. The uncertainties in the risk estimates should also be addressed. This risk analysis process is described in Box 3-3.

PUBLIC/STAKEHOLDER INVOLVEMENT

A challenge facing any Army policy for CAIS disposal is its acceptability to the public.[3] A study by the NRC (1996b, p. 23), cites Fiorino's (1990) approach, under which there are "three compelling rationales" for public involvement—normative, substantive, and instrumental. The normative rationale is that a democratic government should obtain the consent of the governed; citizens have a right to be involved in decisions that affect them. The substantive rationale is that input from diverse public groups, as well as from scientific experts, can provide essential information that improves the quality of a decision. The instrumental rationale is that public involvement may decrease conflict and increase acceptance of, and trust in, governmental agency decisions.

The Army's experience in the Chemical Stockpile Disposal Program, as well as other agencies' experience with hazardous waste disposal, including the evaluation of sites for nuclear waste disposal, has demonstrated the difficulty of implementing policies in the face of strong public opposition. Because public acceptability will certainly affect

---

[3]The terms public and stakeholder, which are used interchangeably, refer to interested individuals and groups at the local and national levels, rather than the general public. Because the collection of original data on public views specific to CAIS disposal was beyond the scope of this study, the committee has heard from only a segment of the interested public and has not conducted an exhaustive survey. Although the committee's statements reflect that limited input, they are also based on the extensive experience of committee members and a body of literature, cited in the text, that documents public opposition to incineration and public views on alternative technologies. This literature indicates the types of issues the Army will have to address in its public involvement program.

> BOX 3-3 Risk Analysis Process
>
> **Identification of Hazards**
>
> The critical hazards for each stage of a disposal option should be identified. That is, the process steps during recovery, initial handling, packaging, transport, storage, disposal and/or treatment, handling of waste materials, long-term storage of CAIS or process wastes, and any other CAIS-related activities should be analyzed for the hazards that may occur during both routine and nonroutine (non-normal) performance of the activities.
>
> **Understanding of Risks**
>
> The risks for each stage must be well understood. It is not necessary to conduct a quantitative risk assessment for each stage, but there should be some indication of how often an exposure to a hazard may occur, or how likely something is to go wrong and what would happen if it did go wrong. Failures under routine and nonroutine conditions should be thought through (e.g., flooding of the storage area, incomplete combustion, helicopter accident, agent monitors improperly calibrated).
>
> **Risk-Control Measures**
>
> Provisions for managing or controlling risks should be examined and risk reduction/mitigation/control measures identified, targeted at key risk factors, and well thought out in terms of feasibility and effectiveness.
>
> **Risk Context**
>
> The risks must be put into context. Risk measures/descriptors should be clear and meaningful to both professionals and lay people. Risk comparisons should not be overly generalized and should clearly state whether the risks refer to all possible CAIS set recoveries or to an individual recovery and have not been scaled up to address the overall problem.

the viability of the Army's proposed policy, it is a key issue that must be addressed through meaningful public involvement that engages the public in developing solutions.[4]

The public, however, is not a single entity. It is composed of many publics—individuals and groups that typically have different criteria for acceptability. In assessing the relative acceptability of policy options, it is necessary to identify the particular publics involved, the issues that are important to them, their ability to influence policy,

---

[4]The NRC recently released a report on alternative (nonincineration) technologies for destruction of assembled chemical weapons (NRC, 1999b) that highlights the problems in interpreting the term "public acceptability." The authors of that report raised questions of who constitutes the public and whether acceptance requires a broad public consensus. They noted that, in controversial programs, there is no single public and that the fragmentation into multiple publics ranging from those who are "engaged" to those who are less interested complicates the problem of gaining acceptance. They concluded that it is generally difficult to reassure these publics that a potentially hazardous facility is safe, especially when trust in the implementing agency is low. In addition, the policy review process offers active opponents of a given course of action or technical approach many opportunities to challenge it. The extended policy debate that ensues both influences and is influenced by the opinions of the multiple publics.

and the potential for addressing and resolving issues that could prevent or seriously delay policy disposal implementation. As described in Box 3-4, a public involvement program for CAIS should identify affected public and stakeholder groups and their key issues, assess their influence on policy, and develop effective means for resolving their key issues.

An effective way for the Army to commit to such a process is through a written, publicly released plan for dialogue with stakeholders. A *written* plan provides several advantages over an unwritten policy:

- It gives stakeholders an opportunity to participate in the development and review of the plan.
- It fosters credibility within the Army for the public involvement program.
- It requires that technical staff and public involvement staff integrate their plans and activities, which is essential if the dialogue with the public is to connect into program decisions.

The plan should be open to the public generally, for example, through public release and broad dissemination.

The key issue affecting public acceptability of a CAIS disposal policy will be the disposal technology proposed for the primary and secondary treatment of process waste. In light of past opposition from concerned groups to incineration in general and to the incineration of chemical weapons in particular, gaining public acceptance for commercial incineration of CAIS is likely to be difficult. Opposition to the incineration of hazardous wastes emerged during the 1980s among a wide range of groups (Curlee et al., 1994; Walsh et al., 1997). Among these groups were local, grass roots groups opposed to having a hazardous waste or municipal waste incinerator in their communities; established environmental groups, such as the Sierra Club, Friends of the Earth, and Greenpeace; and groups opposed to the incineration of chemical weapons in particular, such as the Chemical Weapons Working Group and, in the 1990s, the Non-Stockpile Chemical Weapons Citizens Coalition.

Health-related concerns about emissions and/or formation of dioxins and furans that emerged in the 1980s have expanded and now include questions about the validity of Army estimates of the toxicity of chemical agents. In addition, health and safety concerns have been reinforced by larger concerns than the choice of a disposal technology. Documented concerns include concerns about the extent of public involvement in the decision process, performance and accountability, trust, environmental justice, equity (both geographic equity and the "stigmatization" of communities where hazardous facilities are located), and the future use of facilities (Walsh et al., 1997; Hunter and Leyden, 1995; Curlee et al., 1994; Bradbury et al., 1994; and Rabe, 1994; see also Kasperson et al., 1992; and Edelstein, 1988). Concerns about transportation risks were expressed by members of the Non-Stockpile Chemical Weapons Citizens Coalition during discussions with the committee. As demonstrated in the U.S. Department of Energy's nuclear waste transportation program, concerns about transportation (and handling) risks include concerns about emergency response capabilities at the site and along proposed transportation routes. There is also a general concern and sense of unease about incineration (Hunter and Leyden, 1995). All of these public concerns indicate the difficulty the Army is likely to experience in gaining acceptance for incineration of CAIS, particularly because these concerns would score very strongly for the characteristic of dread identified by Slovic, et al. (1979) as likely to cause problems for public acceptability.

> BOX 3-4 Assessing the Public Acceptability of CAIS Disposal Options
>
> **Identify Affected Publics/Stakeholders**
>
> An effective public and stakeholder involvement program requires first identifying the individuals and groups who believe they will be affected by, or are likely to take an active interest in, CAIS disposal. Among these groups are (1) those with legal, regulatory, or organizational interests (Congress, state legislators, federal and state regulators, branches of the military); (2) those with safety concerns (persons in physical proximity to a disposal or recovery site or a transportation route); (3) those with economic interests (facility owners or businesses and residents concerned about potential impacts on property values); and (4) those with philosophical and ethical concerns (environmental and peace organizations). Some stakeholders may only be interested in particular disposal options.
>
> **Identify the Issues**
>
> Identifying key issues of concern to various stakeholders is a prerequisite for an effective public involvement program and indicates the level of controversy and difficulties the Army may encounter in attempting to implement a particular disposal option.
>
> **Assess Stakeholders Influence on Policy**
>
> Stakeholders can influence policy directly through the regulatory and political processes. The National Environmental Policy Act (NEPA) and state permitting processes, which mandate that opportunities be provided for public review and comment on policy decisions, with associated opportunities for legal review, are the primary mechanisms through which stakeholders can affect policy directly. This influence can result in the imposition of additional safeguards, permit conditions, or regulatory actions that affect the cost and feasibility of policy implementation.
>
> Stakeholders can also affect policy by encouraging state and federal legislators (particularly Congress) to intervene and make changes in the law, which can have a significant, indirect impact on a disposal option by increasing costs and causing delays, hence compromising the economic viability of a proposed option. Networking and coordination among environmental groups, such as the Citizens Clearing House for Hazardous Wastes, have provided resources and advice to communities selected for the siting of unwanted facilities (Szasz, 1994). A similar role is currently being played by the Chemical Weapons Working Group and the Non-Stockpile Chemical Weapons Citizens Coalition.
>
> **Resolve the Issues**
>
> Although the public acceptability of a disposal alternative can only be determined by the participants, a preliminary assessment of issues and stakeholders can indicate how these issues may be resolved. There may be ways of implementing a policy that are more responsive to known stakeholder concerns, particularly by engaging stakeholders in developing potential solutions that are both technically feasible and publicly acceptable.

Although the policy debate was originally focused on opposition to incineration, the focus has now shifted to the development of alternative technologies for the treatment of both primary and secondary wastes. Organized opponents of incineration believe that the federal government has a responsibility to support the development of technologies considered by these groups to be less harmful than incineration to public health (Ginsburg, 1992). NRC and Army reports have documented the opposition of various citizen groups to incineration and their preference for alternative technologies at the chemical stockpile sites (NRC, 1996a; U.S. Army, 1994; Bradbury et al., 1994). For example, responses by the Stockpile Program's Citizen Advisory Commissions to an NRC report, *Recommendations for the Disposal of Chemical Agents and Munitions* (NRC, 1994a), showed that a majority preferred neutralization to incineration. Community stakeholders in Aberdeen, Maryland, and Newport, Indiana, expressed a clear preference for neutralization over incineration of stockpile chemical weapons stored at the sites near them (NRC, 1996a).

Largely because of significant (albeit not unanimous) public opposition to incineration, Congress required that the Army revise its original plans to deploy incineration at all eight chemical stockpile sites in the continental United States. The Army is now developing chemical neutralization processes at both Newport and Aberdeen, where strong community support for neutralization was evidenced during the recent permitting process (Defense Environmental Alert, 1999).

In addition, Congress appropriated funding for the identification and demonstration of at least two alternative technologies under the Assembled Chemical Weapons Assessment (ACWA) Program, while withholding funding for construction of incineration facilities at two other chemical stockpile sites (Blue Grass, Kentucky, and Pueblo, Colorado). An integral feature of the ACWA Program is the establishment of a facilitated dialogue that seeks to integrate the values and perspectives of communities, regulators, and other concerned parties into the process of developing criteria for assessing alternative technologies. The goal of this involvement program is to develop decisions that are publicly acceptable, as well as technically sound.[5]

Although the mobile RRS, which is the baseline technology for CAIS disposal, uses neutralization as the primary technology, a nonincineration technology is not yet available for the disposal of RRS process wastes. The Non-Stockpile Chemical Weapons Citizens Coalition has expressed its desire that the Army use technologies other than incineration for the treatment of secondary wastes.[6] However, it is not known when nonincineration technologies will become available, and there is no consensus among stakeholders on whether it would be preferable to store the wastes in the interim or dispose of them at existing incineration facilities.

PROGRAMMATIC CONSIDERATIONS

Programmatic considerations that affect the choice of a disposal alternative include (1) project scheduling, (2) the availability and sources of funding, and (3) the coordination of the activities of organizations involved in the disposal program.

---

[5] In the NRC report on ACWA (Assembled Chemical Weapons Program), the authoring committee noted that little systematic and reliable data are available on public reactions to alternatives to incineration of chemical agents (NRC, 1999b). However, that committee's discussions with the most actively involved citizen groups suggested that four attributes of a technology were most important to them: (1) the capability of the system to hold and test effluents prior to release; (2) the "transparency" of the technological process; (3) the inclusion of specific plans for decommissioning the facility and remediating the site after all of the stockpile there had been destroyed; and (4) the capability to quickly and safely shut down the facility. Similar criteria, as well as the importance of meaningful public involvement in the selection process, were also expressed by representatives of the Non-Stockpile Chemical Weapons Citizens Coalition in discussions with this committee.

[6] Letter from the Non-Stockpile Chemical Weapons Citizens Coalition to Secretary of Defense William Cohen, November 13, 1998.

Scheduling

The schedule for the destruction of recovered CAIS material is still being formulated. It will not be finalized until the methods of destruction, including viable alternatives to incineration, have been scientifically shown to pose minimal risks, and the necessary permits have been obtained. The Army estimates that using the RRS, all known CAIS items in storage and in known burial sites could be destroyed by the end of the third quarter of fiscal year 2002 (assuming there are no permitting or other delays). The disposal alternative selected will probably be expected to meet a similar disposal schedule. However, additional CAIS items are likely to be found for years to come.

Funding

The Army's preliminary funding projections for CAIS disposal assume that the RRS will be used for disposal and that there will be a specific number of recovery sites. Many questions have still not been answered, however. While the Army has overall responsibility for CAIS disposal, it is uncertain which costs of recovery, transport, and disposal for different disposal alternatives would be borne by the U.S. Army Corps of Engineers, the NSCMP, or the base commander. For example, who will pay the costs of commercial disposal of CAIS items found on an active or inactive military base? Who will bear the cost of on-site storage? Who will bear the cost of CAIS transport? How will long-term CAIS recovery and disposal costs be funded?

Organizations

The destruction of CAIS materials will be based on choices and actions by many individuals and parties who will be both directly and indirectly involved in decisions about locating destruction sites, selecting the methods of destruction, setting schedules, and managing costs. These parties include elected officials at the federal, state, and local levels, as well as federal and state regulatory agencies. Base commanders at active installations and the U.S. Army Corps of Engineers at former military installations, as well as the Army's Technical Escort Unit and others, will be actively involved in the handling, transportation, storage, and disposal of CAIS. The success of the program will depend on how well the staff of the NSCMP addresses the concerns raised by these stakeholders.

4

Review of the Commercial Incineration Option

The Army conducted a preliminary survey of five commercial hazardous waste facilities to gather information on their industrial processes, experience, compliance with environmental regulations, field services offered (e.g., packaging, transportation), public relations, plant safety, costs for services, and business requirements. The technologies used at these facilities include incineration, neutralization, and gas-phase chemical reduction. Because one company asked that it not be identified, none of the five firms was identified to the committee. The Army summarized the results of this survey in its report to Congress and a supplemental technical report (U.S. Army, 1998a; Amr et al., 1998). The committee assessed both reports and the commercial disposal option for CAIS in terms of the "issues to consider" described in Chapter 3. Table 4-1 contains the committee's summary evaluation of the commercial disposal option, assuming that incineration is the technology employed.

TECHNOLOGY

Excerpts from the Army's Report to Congress

3.1 Treatment Technology

This study assessed the capabilities of five commercial facilities using three treatment technologies—incineration, neutralization, and a transportable gas phase chemical (hydrogen-based) reduction system—for CAIS treatment and disposal. Of the five commercial hazardous waste facilities investigated for this study, three would use incineration for CAIS, one would use neutralization, and one would use hydrogen reduction. Hazardous wastes treated by these facilities include agent-contaminated products (e.g., soil), industrial chemicals, pesticides, polychlorinated biphenyls (PCBs), and medical nitrogen mustard (HN). The commercial firm that offers hydrogen reduction has industrial experience treating pure chemical agent, while all the others have treated related materials and other DOD hazardous wastes. It must be noted, however, that the firm offering hydrogen reduction currently has no such plants in operation in the United States and no firm prospects to develop a fixed facility. Their proposed approach for this application would require permitting and setup at each CAIS location, leading to a probable increase in cost over the current RRS-based process.

The ongoing development of non-incineration based alternatives for commercial waste disposal (to include gas phase chemical reduction), under the auspices of the U.S. Army's Assembled Chemical Weapons Assessment (ACWA) Program and elsewhere, must be closely monitored for its applicability to the destruction of CAIS, whether by commercial or government providers. Indeed, the potential for use of residual hardware from the ACWA demonstrations, now underway, appears to provide the most promising path to government-managed operations for CAIS disposal. As ACWA is just moving into the test operations phase, however, it is premature to assess the extent to which this hardware may be of use, or when. The increased public and focus group acceptability of non-incineration based alternatives for CAIS disposal is a factor that must not be underestimated.

TABLE 4-1 Summary Evaluation of the Commercial Incineration Option

	Committee Evaluation
Technology	
Process reliability and effectiveness	Well proven for mustard; arsenic-containing agents may require special treatment.
Technical maturity	Mature but process modifications may be required.
Monitoring and disposal of process effluents	Committee recommends continuous air monitoring in receive/unpack areas; public may require "hold and test" monitoring of emissions and effluents.
Laws and Regulations	
Consistency with present laws, regulations, and treaties	Non-Army disposal requires legal/regulatory relief, clarification, or flexibility; some facility permit modifications may be required by EPA.
Costs	
Permitting	Permit modifications, if required, may add cost; permit restrictions may affect processing costs.
Indemnification	A potential added cost to the Army or the facility.
Facility modifications	Monitoring and other modifications may increase costs.
Transportation	Transportation to commercial sites may increase cost; escorts may be required; not clear who pays for handling, characterization, and transport.
Processing operations	Dedicated processing of CAIS could be costly; CAIS packaging could be an added cost.
Indirect costs	Hidden indirect costs (overhead, administration, maintenance).
Environmental Impacts, Worker/Public Safety, and Risks	
Environmental impact	Air emissions minimized by facility design.
Worker safety	Monitoring in receive/unpack areas needed; added training and protective equipment for handling hazardous waste is needed if not already adequate; hazards seem manageable for facilities permitted for hazardous wastes of comparable toxicity.
Public safety	Impacts on public safety controlled by government regulations.
Risk analysis[a]	Risks generally known and understood for commercial facilities; CAIS chemicals seem similar to other hazardous chemicals currently being incinerated; risks to workers in receive/unpack areas should be analyzed.
Public/Stakeholder Involvement	Perceived public health issue concerning chronic risks from incinerator emissions; transporting large numbers of CAIS, or CAIS types containing large volumes of agent, may be an issue; priority should be on allocating resources for public involvement.
Programmatic Considerations	
Schedule	Could allow prompt disposal of small recoveries of CAIS, but public resistance and regulatory treatment may lead to significant delays.
Funding	Liability and contractual issues could increase costs.
Organizations	Corporate commitment is a significant unknown.

[a]Risk analysis includes identifying hazards, understanding risks, identifying risk control measures, and putting risks into context. The initial discovery of CAIS items, particularly by untrained members of the public, seems to pose the greatest risks. However, the committee's analysis begins at the point of CAIS recovery.

All commercial facilities interviewed for this study can provide field services—such as characterization, packaging, and transportation. In some cases, these services are provided directly by the facility; in other cases, the services are subcontracted or provided through teaming arrangements. Trucking is the primary mode of transportation. Each company also stated that they would be willing to change their procedures in order to support the Army in its effort to treat and dispose of CAIS.

All commercial facilities have secure storage areas and the capability to track and dispose of residues. Secure storage of CAIS for a transportable hydrogen reduction system, however, would rely on Army storage at the CAIS site.

For the relatively infrequent recoveries of CAIS components found packed in metal containers ("PIGs"), continued use of the current RRS system as an unpack, classification, and segregation system would appear to be appropriate from a safety and cost perspective, limiting the need to facilitize unpack areas at contractor facilities. The RRS is increasingly recognized also to have utility for response to terrorist incidents and other chemical emergencies. Costly deployment and field operations of the complex RRS system need not be required for field recoveries of small quantities of exposed glass CAIS components, which can be assessed with portable spectroscopy systems, repacked for safe transportation, and taken to a commercial facility for disposal.

(U.S. Army, 1998a, pp. 7–10)

[L]imited monitoring capability is a potential disadvantage associated with the use of commercial facilities for CAIS treatment. Specifically, emissions monitoring is limited, and none of the facilities interviewed for this study has any agent monitoring capabilities. Although the monitoring for destruction and removal efficiencies (DREs) is not routine, each technology appears to be capable of achieving required DREs. With Army assistance, some companies are willing to install Depot Area Air Monitoring Systems (DAAMS) or Miniature Continuous Air Monitoring System (MINICAMS), if required by Federal law or regulation. Although some process and/or facility modifications (e.g., unpack area) may be required, these issues do not appear to be technically or economically insurmountable.

Ultimately, given the nature of the material involved, and in an attempt to adhere to commercial practices, this study does not recommend continuous air monitoring measures for commercial CAIS destruction. However, portable agent detection systems may be appropriate for use at the time of CAIS delivery to ensure that the containers remain intact after transportation.

(U.S. Army, 1998a, p. 10)

3.2 Engineering Controls

Two of the facilities (both using incineration) have three process lines that could be used for treating CAIS, while the remaining three facilities each have a single process line. All three incineration facilities indicated that they could feed lab packs (i.e., readily available, frequently used commercial containers) and/or drums in their process line. A glovebox is currently being utilized for unpacking at one facility. Most facilities would be able to feed CAIS as a "dedicated" (i.e., CAIS-only) campaign or as a mixed feed with other waste.

The process area of the neutralization treatment technology is located in a closed area that is vented through a scrubber system. Items could be fed individually or in "lab packs" of several vials each. In either case, they would have to be repackaged before being transported to the facility. This company is willing to investigate the possibility of removing the CAIS items from "PIG" overpacks at their facility. Finally, the hydrogen reduction equipment requires little space and could be easily accommodated in a controlled enclosure.

In terms of disadvantages, limited controls were noted for transportation, unpack areas, and process areas. At present, control devices tend to be primarily targeted for liquid spills. Implementation of new controls may not be well received by industry.

(U.S. Army, 1998a, p. 10)

5. Conclusions and Recommendations

1. **At this time there are no known *technical* limitations that would prevent effective destruction of all CAIS materiel in commercial facilities, either incineration or non-incineration based.** Several commercial disposal facilities indicated a capability to destroy CAIS using procedures now in place for materials having comparable or more complex

chemical structure, toxicity, and packaging. These materials include certain pesticides and medical wastes. The feasibility and practicality of implementation, however, needs to be determined. This determination might be accomplished through a case study that considers hypothetical CAIS delivery schedules and a proposed contractual mechanism.

(U.S. Army, 1998a, p. 13)

Committee Evaluation of Technology

The Army's conclusion that technical limitations would not prevent the effective destruction of CAIS materiel in commercial facilities may be correct, especially for sulfur mustard. However, many technical issues must be addressed in more depth, particularly for the arsenic-containing CAIS chemicals, lewisite and adamsite. Sulfur mustard has been destroyed by each of the three technologies surveyed (Amr et al., 1998). Similar technologies are available for the destruction of lewisite, but they are not as fully developed. The technical aspects of incinerating CAIS items are summarized below.

Process Reliability and Effectiveness

Incineration processes have been used extensively for the destruction of sulfur mustard and are likely to be reliable and effective for the disposal of CAIS materials. Sulfur mustard is relatively easy to burn. In the EPA's "incinerability" classification, sulfur mustard is a Class 4 material, which means it is intermediate in combustibility between the easiest (Class 7) and the most difficult (Class 1) hazardous wastes to incinerate.[1] Incineration has been used extensively to destroy sulfur mustard contained in old munitions, with destruction and removal efficiencies (DREs) generally achieving 99.9999 percent, often referred to as a DRE of "six nines" (NRC, 1993). This level of destruction exceeds the level required under international treaties and generally meets state and federal standards under RCRA permit conditions. In four test burns of bulk sulfur mustard at JACADS (Johnston Atoll Chemical Agent Disposal System) in 1992, DREs of 99.9996 percent or better were achieved, significantly exceeding the RCRA requirement of 99.99 percent for this facility (Mitre, 1993). The emissions of products of incomplete combustion were very low—similar to those produced by incineration of fuel oil. The stack emissions of dioxins and furans were also extremely low, 0.0–0.16 $ng/^3$, which is well below the current U.S. standard of 30 ng/m^3 for municipal waste incinerators (NRC, 1994b). EPA has a draft reassessment of health effects of dioxins, but no new standard has been issued.

Based on experience with the incineration of similar chemicals, it seems reasonable to assume that sulfur mustard can be destroyed safely and effectively in commercial incinerators, particularly high-efficiency facilities qualified to burn PCBs. A recent EPA study of the emissions from the JACADS facility detected very low levels (just above the

---

[1] The EPA incinerability index is a semi-empirical guide to choosing appropriate conditions to minimize emissions of the compound being burned, particularly the principal organic hazardous constituent (POHC). EPA (1989) reported that the best correlation with observed destruction efficiency was the thermal stability of a POHC under oxygen-starved conditions. Class 4 on this index indicates that (limited) data on the thermal stability of mustard gas place it as intermediate between the most stable compounds (most difficult to burn completely) and the least stable (highest destruction efficiencies). Experimental data from pilot-scale or full-scale incineration tests that support the index and the POHC rankings are in EPA, 1989, and Lee et al., 1992.

limit of detection) of chlorinated furans and PCBs in about half of the trial burns done under the direction of EPA staff (EPA, 1998a).[2]

Based on laboratory-scale experiments, lewisite appears to be as easy to burn as sulfur mustard (Brooks and Parker, 1979). Although lewisite has only been burned on a modest scale in chemical weapons disposal operations (Petrov et al., 1998), similar arsenical agents such as phenyldichloroarsine are routinely burned on a large scale in the German chemical waste incinerator at Munster (Martens, 1998). However, the arsenic contained in the lewisite is oxidized to volatile arsenic oxides and chlorides (Dempsey and Oppelt, 1993). These oxidation products must be scrubbed or electrostatically precipitated from the flue gases to meet regulatory standards for emissions of heavy metals. Difficulties in meeting regulatory standards for arsenic emissions were a significant factor in the Army's decision to destroy 10 tons of lewisite stored at the Deseret Chemical Depot (DCD) at Tooele, Utah, by neutralization rather than combustion in the existing incinerator at TOCDF (Tooele Chemical Demilitarization Facility).

Some commercial incinerators, however, are reported to have appropriate facilities and regulatory permits for burning arsenic-containing wastes (Brankowitz, 1998).[3] Given the small quantity of lewisite in CAIS (see Table 1-4), it should be feasible to burn lewisite in commercial, arsenic-permitted incinerators. The permits for three commercial incinerators limit the arsenic concentrations in atmospheric emissions to 2–10 ppm, depending on the state in which the facility is located, and the EPA recommends a limit of 3 ppm (Velzy Associates, 1990). Therefore, meeting the regulatory restrictions on the concentration of arsenic in the feed to the incinerator could be a problem. In addition to the restriction on arsenic input in the feed, the EPA also has a rigorous limit of 0.03 milligram per cubic meter (mg/m^3) for even temporary exposure to airborne arsenic (EPA, 1998b).

Reliability and Robustness

A well maintained, PCB-qualified, commercial incinerator with well trained personnel should provide safe, reliable destruction of sulfur mustard and lewisite agents even under nonroutine conditions. Successful incineration of hazardous wastes comparable in toxicity and combustibility to the agents has provided an extensive experience base for operations with agents (Dempsey and Oppelt, 1993). Incineration technology should be adaptable to CAIS samples that are badly degraded, contaminated, or poorly characterized.

---

[2]While this report was undergoing NRC review, a report by a separate NRC committee on the use of carbon filtration for gaseous emissions from stockpile incineration facilities was released (NRC, 1999a). The report on carbon filtration in which the EPA trial burn data for JACADS and TOCDF were reviewed, contains the following finding:

Finding 1a. The reported emitted concentrations of SOPCs [substances of potential concern] measured during trial burns at the JACADS and TOCDF incinerators are among the lowest reported to the EPA. TOCDF emissions are the lowest, or at least one of the lowest, in dioxins, mercury, cadmium, lead, arsenic, beryllium, and chromium. The reported emissions of some SOPCs were based on the analytical detection limit for the constituent, which means the actual concentration could be much lower than the reported concentration....

[3]According to verbal information given to committee staff by EPA staff, all or most of the commercial hazardous waste incinerators listed in Table 2-1 have permits for handling arsenic-containing wastes.

Handling and Decontamination of Containers

Most commercial incinerators can handle wastes packaged in a variety of ways. If the outer packing is not too large, the entire package may be fed directly to the incinerator. This procedure would reduce risks to personnel who might be exposed to agent vapor when handling leaking, fragile, or badly corroded containers. Combustible packing materials, such as wood, paper, or sawdust, would be destroyed along with any agent adsorbed on the packing. The heat of the incineration process would decontaminate noncombustible materials, including glass and metal.

Technical Maturity

Incineration has been used extensively to destroy sulfur mustard from both bulk containers and munitions. Experience with large-scale incineration of HD in Canada, Germany, Iraq, the United Kingdom, and the United States (JACADS and Rocky Mountain Arsenal) was summarized in a 1993 NRC report. DREs generally exceeded 99.9999 percent.

Monitoring and Disposal of Process Effluents

Monitoring issues in CAIS disposal operations arise primarily at the beginning and end of the process. At the beginning, when CAIS containers arrive at a facility and are put into the disposal system, the major concern is agent vapor that may leak from damaged packages or may be released when a container is opened. These concerns are common to all destruction technologies but are of most concern in a neutralization process. With incineration, for example, packaged CAIS items may be introduced directly into the incinerator or reactor. Neutralization may require that the containers be opened in a glove box, as is planned for the RRS operation (Brankowitz, 1998). However, in all these cases, continuous air monitoring for CAIS agents should be used in the receiving/unpacking area to ensure that workers are not exposed to unsafe levels of agent vapor.

The Automatic Chemical Agent Monitoring System (ACAMS) is used for monitoring workplace air in chemical stockpile disposal facilities. A limitation of the current ACAMS is that it is designed to monitor the concentration of a single agent, such as HD. The newer Miniature Continuous Air Monitoring System (MINICAMS) is designed to monitor more than one agent but may not work with lewisite. Another limitation that could cause problems at a commercial incinerator is that the monitors are subject to interference by chemicals other than the agent. In a commercial facility, which may handle hundreds of different chemicals, interference could cause false alarms that would interrupt the operation of the facility. Commercial operators may be reluctant to add air-monitoring capabilities beyond the ones already in place. Despite these issues and obstacles, which will require further consideration, the committee believes, as noted above, that continuous air monitoring for CAIS agents should be required in the receiving and unpacking area when CAIS are being handled.

Monitoring at the end of a disposal process is intended to ensure that hazardous materials are not released into the environment. The form of monitoring depends on the nature of the process. For incineration processes, the major concern is gaseous effluents, although liquid or solid residues must also be disposed of safely. One problem with gaseous emissions is that retaining the gases until they have been analyzed and certified safe is not practical. In fact, most commercial incinerators do not routinely monitor flue gases for the presence of the material being burned, except during trial burns when effluents are analyzed to show that the concentrations of unburned waste, products of incomplete combustion, and particulate matter do not exceed the levels specified in the facility's permit.

A particular problem for burning arsenic-containing materials, such as lewisite and adamsite, is the difficulty of monitoring arsenic emissions in stack gases. In tests of the EPA's Combustion Research Facility in Arkansas, most of the arsenic emitted was not detectable by the analytical methods then in use (Lee et al., 1987). With chloride-rich feed streams, such as lewisite, the emissions will probably be a mixture of volatile oxides, chlorides, and oxychlorides of arsenic (Dempsey and Oppelt, 1993).

The liquid or solid wastes produced from incineration or any of the destruction technologies surveyed for the Army (Amr et al., 1998) could be held for analysis before disposal. Solid wastes from hazardous incineration are usually sent to a hazardous waste landfill. Unless CAIS materiel is burned separately from commercial waste, the combustion residues are not likely to be monitored.

LAWS AND REGULATIONS

Excerpts from the Army's Report to Congress

3.6 Facility-Specific Regulatory Issues

There are several regulatory issues relevant to the use of commercial facilities for CAIS treatment and disposal: (1) CAIS interstate transportation, (2) facility permitting requirements, and (3) public notification procedures. Most of these can be simplified considerably by amendment to Federal law. Others can be resolved only subsequent to changes in the Federal law through detailed negotiations with State regulators, commercial providers, and key citizens' groups. Due to time restrictions in this study, and the desire to determine first if there was enough commercial interest to warrant further progress on this subject, substantive discussions with Tribal, State, and Federal regulators have not yet occurred. The existence and conduct of this study have been briefed to and discussed with members of the public and Tribal, State, and Federal regulators in open forums, however, and venues are open for further discussions following this preliminary study.

As stated above, with the exception of the hydrogen reduction system, all commercial facilities in this assessment are currently permitted (the hydrogen reduction system was permitted in a previous operation). Moreover, these same four facilities are currently permitted to process RCRA waste codes that might allow the receipt of CAIS. For example, the RCRA waste codes for nitrogen mustard on the RRS permit application for the state of Utah are D004-D011, D022, and P999. If these specific "D" waste codes were to be applied to CAIS in the states where the facilities are located, CAIS might be accepted by those facilities interviewed for this study. It is presently indeterminate as to whether permit modifications would be required for the treatment of CAIS.

(U.S. Army 1998a, p. 12)

5. Conclusions and Recommendations

. . .

2. **The law and its interpretation are the major obstacles limiting options for the transportation and disposal of CAIS.** Amendment of applicable Public Law is needed because it drives the rigorous compliance requirements governing the Army. CAIS materiel is demonstrably different from other chemical materiel: (1) CAIS contains no nerve agent; (2) CAIS containers are glass, and easily accessed; (3) CAIS were intended for training purposes and never intended for lethal purposes; (4) CAIS are not associated with explosives; (5) CAIS contents are not much different in toxicity than the industrial chemicals handled on a routine basis by industry and by commercial hazardous waste facilities; and (6) CAIS consist of relatively very small quantities of materiel (often just a few ounces). On these grounds, a

strong case may be made that CAIS should be exempted from the provisions that govern chemical agent stockpile demilitarization.

3. **Strict application of the laws and regulations that were specifically tailored for chemical weapons storage and disposal operations may not be applicable for disposal of CAIS and in the presence of these provisions might deter commercial facilities from participating in the disposal of CAIS. Notwithstanding any changes in the legal and regulatory status of CAIS, however, CAIS should continue to be centrally managed by the Army as non-stockpile chemical materiel to ensure the application of otherwise appropriate criteria.** Even if the extraordinary oversight currently provided by the Army for CAIS destruction were relaxed, the use of commercial firms does not mean Army "hands-off." RCRA hazardous waste requirements and government oversight would still apply. Acceptable Army criteria (e.g., safety, engineering controls, monitoring, security, public affairs) beyond commercial practices should be defined, and data should be collected for ensuring compliance with these criteria. The definition and execution of confidence methods which the Army could apply to ensure public health and safety and protection of the environment during commercial destruction of CAIS needs to be developed. For example, the Army could provide government supervision of the disposal action to ensure total destruction (using procedures similar to those now in use for drug contraband destruction at commercial facilities). Requirements could also be placed on commercial disposal facilities to process CAIS materiel with other wastes to ensure an optimum feed mix as further insurance for safe destruction.

(U.S. Army, 1998a, pp. 13–14)

Committee Evaluation of Legal and Regulatory Issues

A viable commercial incineration option for CAIS disposal will require changes, clarifications, or flexibility in existing laws and regulations. A number of regulations and legal interpretations now mandate that the Army maintain control of CAIS materials during transportation and disposal. Even if the transport and disposal of CAIS can be accomplished without Army facilities or personnel, regulators may require modifications to a commercial firm's operating permit, depending on the disposal classification assigned to CAIS components. Because a case-by-case determination at every site would not be cost effective, the Army and EPA could develop a presumptive CAIS treatment and storage guideline, a permit-by-rule, a national permit, or some other method of establishing standard conditions that could be used for evaluating any site that might handle CAIS.

COSTS

Excerpts from the Army's Report to Congress

3.7 Cost for Services

In light of the recurring costs of permitting, transportation, set-up, and closure for each site where CAIS are located, the potential for generating cost savings by treating and disposing of CAIS in commercial facilities may be considerable. Prices for commercial destruction of CAIS-type material can be as low as a few dollars per pound, before the costs associated with the additional requirements identified in this report are imposed. While government equipment such as portable spectroscopy systems may be required at all burial remediation sites as appropriate due to the need to characterize and re-package CAIS components prior to shipment, the use of commercial facilities for ultimate destruction still offers substantial cost benefits.

The use of commercial facilities also could lower costs by reducing the need for multiple investments in Government equipment, when capability is needed for concurrent destruction activities at multiple sites. Accessibility to some of the facilities interviewed for this study may be easily acquired from existing contract mechanisms. The disadvantages identified during site visits with the commercial facilities included (1) cost-reimbursable funding for plant modifications may be required, and (2) open competition is not desirable (but may not preclude some from participating).

(U.S. Army, 1998a, pp. 12–13)

5. Conclusions and Recommendations

. . .

4. **If current Army practices for managing the disposal of CAIS as chemical agent are changed, considerable cost savings might be realized, given that commercial facilities and technologies are available today for CAIS treatment and disposal.** The magnitude of these savings has yet to be quantified definitively. However, the cost of an RRS deployment and limited operations may amount to $2 million or more, so that an RRS operation to destroy a few CAIS vials may cost up to several hundred thousand dollars per vial. While comparable toxic wastes are typically destroyed for a few dollars per vial, those costs do not include the potential costs related to additional requirements outlined in this report. It is recommended that the commercial cost component be further established. A facility pricing structure (e.g., start-up costs, per unit prices) should be developed, and a comparative cost analysis of commercial facilities with the RRS and various RRS scenarios should be conducted. Once any outstanding issues are resolved, the Army may choose to test implementation (e.g., consider an indefinite delivery/indefinite quantity type of contract) before changing its programmatic strategy for destroying CAIS.

(U.S. Army, 1998a, pp. 14–15)

Committee Evaluation of Costs

In the Army's assessment of the viability of processing CAIS materials commercially, interviews were conducted with five companies with the technical capability of processing CAIS items (Amr et al., 1998, p. 15). Many issues involving costs were not discussed. For example, the interviews with commercial firms did not address the costs of adding agent monitors and alarms for detecting agent leaks, costs of training workers to handle agent, costs of personal protective equipment and decontaminating it, costs of a public involvement program, costs of obtaining permit modifications to allow CAIS to be received and destroyed, reporting costs to state regulatory agencies, and possible costs of plant modifications, such as an unpack area for receiving CAIS. The companies appeared to consider the processing of CAIS chemicals as comparable in cost to the disposal of other reactive hazardous wastes that they were permitted to process. In the 1998 report to Congress, the Army acknowledged that the general subject of costs was not fully considered in the report and concluded that "a comparative cost analysis of commercial facilities with the RRS and various RRS scenarios should be conducted. In the Army's report, the costs of commercial facilities that handle very hazardous materials, such as PCBs, were not evaluated. Such facilities would provide a more realistic comparison with CAIS disposal and may have significantly higher costs than those surveyed by the Army.

In the discussion that follows, the estimated commercial disposal costs at the surveyed facilities are compared with a 1997 Army cost estimate for a proposed use of the RRS. The committee was given an Army report estimating the costs for a proposal to dispose of recently recovered CAIS materials (i.e., seven PIGs) at Fort Richardson, Alaska. In the estimate, the CAIS would be sent to DCD for disposal in the RRS. The Army estimated the total direct and indirect capital cost to be $1,784,429 and added a

30 percent contingency, which resulted in a total project cost of $2,319,758 (U.S. Army, 1997a, p. 3-34). In contrast, four of the five commercial disposal firms interviewed provided cost estimates for processing the same materiel brought from Fort Richardson to their facilities. These estimates ranged from $27,633 to $34,878, including transportation, packaging, and processing (the fifth firm gave a lump sum estimate of $1,500,000). The disparity between the commercial estimates and the Army's estimate raises a number of questions about the assumptions used to arrive at these figures and about CAIS disposal cost issues.

Permitting

The costs of obtaining modifications to existing permits for commercial firms considering CAIS disposal were not quantified in the Army's report to Congress. All of the interviewed firms expressed a desire that, as the waste generator, the Army classify the CAIS as a waste that they were already permitted to handle (i.e., D003 nonexplosive, reactive waste) to avoid having to modify their RCRA permits. If a modification were required, the estimated time varied from 90 days for a minor modification to a year or more for a major modification. The costs of permit modifications and associated costs arising from legal challenges to allowing the firms to dispose of CAIS were not estimated.

In contrast, the Army's cost analysis assumed that a RCRA permit would cost $250,000; that compliance with requirements of the National Environmental Policy Act (e.g., conducting an environmental assessment) would cost another $75,000; and that the costs of other permits, fees, and taxes would be about $112,000. Although not all of these costs would be incurred by commercial firms, a realistic estimate of the costs, time, and hurdles involved in obtaining permit modifications should be included in future cost analyses of the commercial alternative.

Transportation of CAIS to a Commercial Facility

Assuming that the commercial transport of CAIS from a discovery site to a commercial facility is possible and that the carrier is not required by permit or law to follow unique procedures, with substantial cost implications, for transporting hazardous materials, transportation costs should not be a major component of CAIS disposal costs for a commercial firm. The issue of who would incur the cost of transporting CAIS materials from the discovery site to the commercial disposal facility (the Army, a commercial carrier, or the disposal firm) is open, although in the interviews, it was assumed that transportation costs would be borne by the commercial facility.

The interviewed firms estimated the cost for picking up materials from Fort Richardson ($100 per pickup) and transportating them to the disposal facility. (Transportation costs varied from $2.00 to $2.85 per mile for truck transport of all CAIS materials in a single shipment from the storage site to the TSDF, accompanied by one field technician.) Estimated transportation costs, using military transport, were three times higher in the Army's cost estimate for the disposal of CAIS from Richardson, Alaska, to an RRS at DCD in Utah. Because these costs differ by a factor of three, the committee questions the realism in the estimates.

Packaging of CAIS Items

The cost of packaging CAIS items at the discovery site was estimated by only one of the five firms interviewed. This estimate was then used for three of the other four firms (the fifth firm gave a lump sum estimate for the entire job). The packaging cost estimate ($14,828) accounts for about half of the total estimated cost of CAIS disposal for four of the five firms. It is not clear whether the packaging of CAIS items in the field would be done by the commercial firm or by Army personnel. If the CAIS items were characterized in the field, sorted into industrial chemicals and chemical agents, and then placed into "labpacks" or other overpacks by Army personnel, and if these costs were borne by the Army, this would reduce the costs of processing by commercial firms. The general question of when a commercial firm's responsibilities and costs would begin (upon CAIS discovery, at the discovery site following Army characterization and packaging, or at the plant gate) should be clarified in future cost comparisons and cost/risk trade-off studies. Commercial cost estimates that assume packaging or other cost elements will be borne by the Army should be supplemented by realistic estimates of the Army's cost for these items so that valid comparisons can be made.

Processing Operations: Facility Modifications

The nature of the CAIS materials received by a commercial firm will have a bearing on the processing costs. If the CAIS items have been characterized, sorted, and placed in appropriate overpacks by Army personnel in the field, the costs of characterization and materials handling at the commercial facility will be lower than if the facility is required to receive the CAIS items "as found." For example, if the facility is required to build an unpack area, perhaps including a device (e.g., portable isotopic neutron spectroscopy [PINS]) for agent identification and a glove box for opening CAIS overpacks and sorting CAIS items, the processing costs would certainly increase. If the facility receives presorted CAIS items that have already been characterized and if these items could be placed directly into the facility's processing unit (e.g., an incinerator or a neutralization chamber) with a minimum of handling or storage, the costs of materials handling would certainly be reduced.

The costs of other facility modifications were not addressed directly by the commercial firms. Modifications could include adding waste handling areas for solid or liquid wastes from CAIS processing; installing and maintaining monitors to indicate the presence and quantities of any CAIS material that escapes engineering controls; and adding emissions monitoring equipment to check for products of incomplete combustion that are not otherwise monitored. Other facility-related costs could include training staff to handle CAIS items and adding plant security measures.

Processing Operations: Direct Costs

Estimates of disposal costs for CAIS items were made by the four commercial firms that provided itemized costs during the survey. The disposal costs for CAIS items (seven PIGs) brought from Fort Richardson, Alaska, to the commercial firms ranged from $225 to $1,428 per PIG. The bases of these estimates were not provided. Treatment/disposal costs for the same CAIS items brought to an RRS located at the Army's DCD in Tooele, Utah, were provided in Attachment F-1-10 of the Army's cost estimate (U.S. Army, 1997a). The Army's labor costs over a 13 workday period of processing operations were estimated to be $169,472, or $24,210 per PIG. This cost estimate assumed that 13 staff members would be required to operate the RRS and that additional staff would be provided by DCD.

In addition to processing operations, RRS labor costs at DCD included mobilization and site preparation, set-up, operations, closure, demobilization, and site cleanup. When these costs were included, the total labor cost for disposal of the seven PIGs was $452,352, or $64,622 per PIG.

The reasons for the very large difference between commercial and Army costs for processing the same quantity of material are not explained in the Army's report. Further discussions with commercial operators will be necessary to ensure that the underlying assumptions about the number of staff required are realistic. The same point applies to the Army's estimates. For example, the Army assumes a labor cost of $64 per hour for all staff, from the RRS supervisor to security guards. A more careful estimate of unit labor costs and the number of staff required may result in lower estimates.

The costs of materials and equipment (e.g., laboratory supplies, decontamination supplies, forklifts, personal protective equipment, utilities, waste containers, and waste disposal), can add to the costs of CAIS disposal. Only one of the five companies interviewed provided an estimate of material and supply costs: $568 (for Vermiculite, 55-gallon drums, and protective equipment). In contrast, the Army's cost estimate for materials and equipment, not including transport and storage of CAIS and equipment usage fees, was $224,693. Although most of the RRS-associated materials costs may not apply to disposal in a commercial facility, further discussions should be held with commercial firms to determine the nature of the materials, equipment, and supplies required to dispose of CAIS items.

Processing Operations: Indirect Costs

The treatment and magnitude of indirect costs varied greatly between the commercial and the RRS options. Indirect costs (e.g., for commercial facility management, administration, preparation of plans, and general overhead) were not provided by any of the firms interviewed. In contrast, the Army assumed that engineering and management costs would be 20 percent of direct labor, materials, equipment, and travel costs. For CAIS disposal at DCD, these costs were estimated to be $224,528.

Processing Rates

Although cost is not directly proportional to the processing rate, the processing rate can affect the economic desirability of commercial processing. The commercial firms interviewed indicated that if CAIS were categorized as a hazardous waste that they were currently permitted to handle, they could commingle the CAIS items with other similar wastes; thus the cost of processing the CAIS would be negligible. It may well be, however, that the chemical agents in CAIS would have to be processed separately, which would result in a substantial underutilization of commercial incinerators or other disposal equipment designed to process larger quantities of material. For example, although the RCRA permit issued by the state of Utah for RRS processing of CAIS items allows the liquid neutralization wastes (containing up to 50 ppm agent) to be disposed of in commercial incinerators, the permit requires that these wastes be disposed of separately from other hazardous wastes (i.e., commingling is not allowed). If this permit restriction is representative of permits issued in other states and is extended to the commercial processing of the CAIS chemicals themselves, the dedication of commercial facilities to CAIS items may result in higher disposal costs than were estimated. This possibility should be explored in further discussions with commercial hazardous waste disposal firms.

Recovery of Overhead and Development Costs

In the interviews conducted with commercial firms, the costs of the design, engineering, fabrication, and upkeep of disposal facilities were not included in the cost estimates. Compared with the quantities of commercial hazardous wastes processed by these facilities, the costs for the very small quantities of CAIS materials would be insignificant. In its cost analysis, the Army did include these costs as a "usage fee" for the RRS itself and for an associated mobile laboratory. The usage fees, based on equipment design and fabrication costs, maintenance and replacement costs, spare parts, and operator training requirements, were $5,100 per calendar day for the RRS and $2,150 per calendar day for the mobile laboratory. For the 47 calendar day period of RRS operations at DCD, usage fees of $326,250 were listed under "Materials and Equipment" in Attachment F-1-10 in the Army's cost estimate. In future discussions with commercial firms, the issue of cost recovery should be explored.

ENVIRONMENTAL IMPACTS, WORKER/PUBLIC SAFETY, AND RISKS

Excerpts from the Army's Report to Congress

3.3 Safety and Security

All facilities have procedures in place for a wide variety of highly hazardous material, and each facility complies with the Occupational Safety and Health Act (OSHA). These commercial facilities operate within the safety requirements continuously and daily throughout the year. OSHA Level A protective clothing is available for use at each location. Personnel at each of the facilities receive OSHA and RCRA-mandated hazardous waste operations and emergency response ("HAZWOPER") and other training. Spill response and contingency plans are also in place at each facility.

For the most part, the facilities interviewed in this study tend not to be familiar with "chemical warfare agents" as such. However, some routine operations are carried out with chemicals having toxicity comparable to or greater than that of mustard and lewisite. Security measures may have to be increased in some cases.

(U.S. Army, 1998a, p. 11)

The report did not contain any conclusions or recommendations specifically addressing environmental impact, worker/public safety, or risk-related issues.

Committee Evaluation of Environmental Impact, Worker/Public Safety, and Risks

The Army must consider the environmental impacts of the routine or accidental release of gaseous or liquid waste streams in its plans. Because commercial incineration facilities must conduct trial burns before being granted a RCRA operating permit, any air emissions should be minimal and safe by design, assuming that CAIS are adequately destroyed under the permitted operating conditions.

Workers at commercial incineration facilities routinely handle hazardous materials. The Occupational Safety and Health Administration (OSHA) regulates worker safety at commercial facilities. (Army regulations, which may in fact conform to OSHA rules, are used at Army disposal facilities.) Given the experience of these commercial facilities, worker safety issues seem generally manageable. However, workers would be unaccustomed to working with sulfur mustard and lewisite, which pose unique handling hazards. Therefore, unless CAIS can be fed directly into an incinerator furnace without being unpacked, special worker training or protective equipment may be required.

No risk assessments of commercial facilities were included in the Army's report to Congress based on a systematic risk assessment process (see Box 3-3). However, because existing commercial facilities would be used, it is reasonable to assume that the risks of handling highly hazardous materials are already well known, well understood, and accounted for in the basic operations of these facilities. Depending on the facility, some hazards associated specifically with CAIS chemicals, such as the handling hazards mentioned above, may require additional risk analysis and facility preparation. The Army's survey, which was limited to a small number of commercial facilities, does not appear to take into account the full range of safeguards that might be available at facilities approved for wastes with high-contact hazards, such as PCBs, dioxin, or even medical waste. Facilities prepared for these hazards may already be prepared to handle CAIS hazards. The risks to commercial disposal personnel from unpacking CAIS items from their transport containers apparently were not addressed.

The chance of something going wrong at any stage of the operation was not taken into account in the technical or cost evaluations—two areas where this issue might have been addressed implicitly, even if it was not discussed explicitly as a "risk." The types of things that could go wrong include accidents in transit or in the handling of CAIS items, changes in ownership of a commercial facility, changes in regulations or community opposition to continued receipt of CAIS materiel.

It can be fairly assumed that most of the risks, other than for recovery, are not unique to CAIS items and that they are well understood by someone (e.g., commercial firms, the Technical Escort Unit), if not by the Army. However, one would expect that the Army would want to identify and evaluate these risks before selecting a course of action, especially if Army personnel will not be responsible for the transportation of the CAIS items (it appears that the Army would be responsible for recovery and packaging in any event).

Because the Army has not specified requirements for monitoring or physical security during storage and handling, the question of how well risks at these stages would be controlled is still open. For instance, because the Army plans to send small shipments and to process them immediately upon arrival, no provisions were specified for storing CAIS items in case of regulatory or other delays.

The Army has stated that CAIS materials are similar to hazardous materials that are routinely handled by commercial disposal facilities. Although some materials are as hazardous as the chemicals found in CAIS, many of them are much less hazardous and occur in more benign forms (e.g., solids, dilute solutions, etc.). Thus the Army's statement could be misleading. By not addressing the issue of risks during the discovery stage, the Army may be conveying a false impression of having the entire situation under control and posing virtually no risk to the public.

PUBLIC/STAKEHOLDER INVOLVEMENT

Excerpts from the Army's Report to Congress

3.4 Public Affairs

The treatment technologies in this preliminary assessment include alternatives to incineration, which are expected to be more acceptable than incineration to the public in general, and non-incineration focus groups in particular. The choice of technology ultimately selected is expected to be very sensitive regardless of the comparatively very small quantities of materiel involved in CAIS disposal.

All the facilities [surveyed by the Army] claim good relationships with their local communities, a situation that they do not want to disrupt. The transportable hydrogen reduction system was permitted during past operations. All other facilities are currently permitted (RCRA Part B and others) and operating. All facilities have experience processing chemicals with toxicity similar to or greater than the chemicals in CAIS and operate 365 days per year.

In general, the facilities want to limit their public outreach efforts. If public notification is required, some facilities will withdraw from consideration. If a CAIS disposal contractor is solicited using a full and open competition, one company will withdraw from this study and not participate in the solicitation. One company, however, was proactive and stated that it would notify the public even if no permit modification were needed. Another company indicated that it would discuss treating CAIS with its employees before making a commitment to the Army.

It is clear that the willingness of candidate service providers to engage in meaningful public outreach and involvement to the standards desired by the [Project Manager for Non-Stockpile Chemical Materiel] PM NSCM will be a major factor in the ultimate acceptability of commercial disposal alternatives in general, and in the source selection of individual service providers in particular. The DOD will work closely with the local community through Citizen Advisory Committees to assure them that the DOD, its agents, or contractors will dispose of CAIS in an environmentally safe and effective manner.

3.5 Corporate Commitment

The initial responses from corporate officers are very favorable. Some expressed the opinion that CAIS treatment involves standard operating practices, given the types of hazardous industrial chemicals with which they already deal. A disadvantage to the Army, as the waste generator, is that government control of the residual waste streams may be limited. Furthermore, legal ramifications (particularly, liability and indemnification) may arise when contract issues are discussed.

(U.S. Army, 1998a, pp. 11–12)

5. Conclusions and Recommendations

. . .

4. **The Army's public notification process is a potential deterrent for the commercial hazardous waste facilities that participated in this study.** A public outreach strategy needs to be developed and mutually agreed upon between the Army, citizens' representatives, and the commercial hazardous waste facilities. Selection of non-incineration based technologies could be expected to reduce the likelihood of public and/or focus group concerns with commercial CAIS disposal.

(U.S. Army, 1998a, p. 15)

Committee Evaluation of Public/Stakeholder Involvement

Public acceptability will affect schedule, cost, and ultimately the viability of the Army's preferred option for CAIS disposal. As discussed in Chapter 3, there are normative, substantive, and instrumental reasons for the Army to place a priority on developing a public involvement program that engages the various public and stakeholder groups in both addressing the issues of concern to these groups and seeking solutions that are technically feasible and publicly acceptable.

The Army's report is focused on projected costs, technical efficiency, and legal issues. The report includes a limited discussion of public involvement, rather than an analysis of the elements of public acceptability. It does not mention the substantial steps taken by the Army over the past two years to expedite public involvement activities for the NSCMP in general. These activities include initiation of a dialogue with citizens' groups and an analysis of issues of tribal and environmental justice. However, the report does acknowledge that public acceptance is one of the issues that must be resolved before a final determination is made to pursue the commercial disposal option. It also notes that "in general, the facilities are concerned about their public outreach efforts," and "if public notification is required, some TSDFs will withdraw from consideration."

The report does not indicate how the proposed commercial disposal option is linked to broader, long-term program goals. Army staff have made considerable efforts to establish constructive relationships with several public groups, including members of the Chemical Weapons Working Group and the Non-Stockpile Chemical Weapons Citizens Coalition, which is the chief organization opposing incineration. A key consideration for the Army is to maintain and build on the current level of trust with these groups, as well as to develop working relationships with a broad range of groups interested in and affected by CAIS disposal. The attitudes toward specific disposal options within this diverse "public" are likely to range from varying degrees of conditional acceptance, even approval, to varying degrees of dislike, including strong, vocal opposition. The actions taken by the Army in pursuit of the CAIS commercial disposal option will certainly be evaluated by stakeholders and will affect other aspects of the program that could, in turn, affect the Army's ability to achieve its broader goals.

In light of these implications for both CAIS disposal and the larger program, the Army should, first of all, accept the fact that public involvement, and not merely public notification, is a necessity whether or not it is a legal requirement. The committee recommends that the Army adopt and consistently adhere to a policy of discussing plans for using a commercial facility with the broader public in the affected community at a very early stage. Second, the committee recommends that the Army initiate a national stakeholder process to provide input for a decision on whether to dispose of CAIS by commercial incineration. Input would be solicited from stakeholders on their views of the risks in disposing of CAIS at commercial incinerators, ways to evaluate risks, and the factors to be incorporated in the substantive criteria for deciding whether the risks of a disposal option are acceptable. Third, the substantive decisions should incorporate and reflect this dialogue. A report summarizing the views of the many stakeholders (including representatives from the locations of commercial incinerators, organized groups concerned about chemical weapons disposal, the states, national environmental groups, industry, and groups near sites from which CAIS will be removed) would be part of the Army record of decision.

How might such a policy be implemented, given that formal public notification may not currently be a legal requirement? At a minimum, before the first time that the Army sends CAIS to a commercial incinerator, the Army should notify the public. The state and the Army should also hold a public meeting. If a permit modification is required, this public meeting could be part of the public notification procedures for modifying the incinerator's permit. Even if no modification is required and the Army's risk evaluation concludes that the incremental risk from CAIS disposal at a facility will be minimal, the public notification process should still be followed before the first shipment to a facility. The committee believes that notification of Congress and the Centers for Disease Control and Prevention is unnecessary for CAIS disposal. But a *community* notification process, as suggested above, will be essential for a public involvement program to be credible.

Although it was not possible for the committee to gather original data for this study, the Army's previous experience indicates that chemical weapons disposal activities are highly visible and subject to broad media coverage. The public will certainly scrutinize any policy that is proposed, and no policy will have the immediate support of all public groups. Therefore, extensive public involvement will be necessary. Effective public involvement is based on the identification of the various "publics," or stakeholders (individuals and groups interested in and affected by a policy), the issues important to each group, and the opportunities available to them to influence policy. A public involvement program provides an indication of public acceptance, and hence the viability, of a policy, as well as ways to negotiate solutions. The Army has begun this process by establishing a mechanism for dialogue among Army staff, public interest groups, and affected communities. It should ensure that its CAIS disposal program has adequate resources for public involvement—sufficient funding and staff with the necessary skills and experience.

Based on the experience of the Chemical Stockpile Disposal Program and similar programs (see Chapter 3), none of the options for CAIS disposal is likely to be entirely acceptable to all stakeholders. Experience suggests that commercial incineration may be the least acceptable option. It may also be the most sensitive to public opposition because public opposition can affect the profits and even the continued operation of a commercial facility. Only a limited number of facilities are capable of achieving the required DREs. As noted in the Army's report to Congress, the operators of some of these facilities may be unwilling to accept the Army's business if they fear that public notification about CAIS being treated in their facilities could provoke public opposition, not just to CAIS disposal but also to the continued presence of the facility in the community. Yet, even the RRS process, which is the baseline disposal method, currently assumes that secondary wastes will be incinerated, which will probably be opposed by the same groups opposed to commercial incineration.

Key Stakeholders

Although a range of individuals and groups could be affected by the Army's CAIS disposal policy, not all of them will consider themselves affected or will take an active interest. Stakeholders are those who *perceive* themselves to be affected and are therefore likely to take an active interest in a program or policy. Stakeholders may have legal or organizational responsibilities, physical proximity, economic interest, or environmental or philosophical interests. Groups with organizational and legal responsibilities include different branches of the Army, Congress, and federal and state regulators. Commercial disposal facility owners will be interested for economic reasons. Populations near an existing commercial disposal facility, near a site where CAIS are found or currently stored, or along proposed transportation routes will see themselves as affected because of their proximity. Groups with an environmental and philosophical interest include the Non-Stockpile Chemical Weapons Citizens Coalition and the Chemical Weapons Working Group, as well as more broadly based national and regional environmental groups, such as the Sierra Club and Greenpeace.

This section reviews the commercial disposal option based on available information about stakeholders and the issues that concern them. Although this review provides a preliminary account of some key stakeholders and their likely associated issues, the committee emphasizes that here, as in any policy debate, *public acceptability can only be determined by the participants themselves through a process of public involvement.*

Key Issues for Each Stakeholder Group

Army. Key issues for the Army are schedule, cost, technical efficiency, safety, political feasibility, and the certainty and clarity of a defined path forward. Different branches of the Army are actually different internal stakeholders with different interests, which may pose difficulties for the NSCMP in implementing its preferred policy. For example, the Chemical Stockpile Disposal Program (which is banned by law from disposing of non-stockpile or other wastes in a stockpile disposal facility) is interested in avoiding any discussion related to non-stockpile issues that could destabilize its program. Similarly, commanders of active military bases are likely to be concerned about storing non-stockpile wastes or having their bases become a storage or disposal site, if these actions result in cost and controversy for their priority missions.

Congress. Congress' desire to limit costs and reach a quick policy decision is likely to be frustrated by the public visibility and public controversy of the issue of CAIS disposal. As the Chemical Stockpile Disposal Program has demonstrated, strong stakeholder concerns are likely to result in delays and even mandated changes in the program. Concerned constituents may be (1) near a commercial facility and opposed to the use of the facility for types of waste not originally included in the permit; (2) near a facility where residual wastes may be stored if incineration is not used; (3) located along transportation routes (assuming that transportation of CAIS is legally permissible) and concerned about transportation risks and local emergency response capabilities; or (4) members of interest groups opposed to the incineration of either primary or secondary wastes.

State Legislators and Regulators. The states, which will be responsible for issuing modifications to existing permits for commercial disposal facilities, will have a significant effect on the viability of the commercial disposal option. Like Congress, the states will be strongly influenced by constituents' views and concerns. Because these views are likely to differ, their impact will depend on the political strength of the stakeholders and their representatives.

Local and Tribal Populations. Local and tribal populations can be categorized by location: near CAIS discovery sites; adjacent to transportation routes (if transportation of CAIS by private companies is legally acceptable); and near a commercial disposal facility. In general, people located near CAIS discovery sites are likely to favor early removal and either storage or off-site disposal, regardless of facility type. This is particularly true if CAIS are found in, or close to, residential areas rather than at existing military sites. People along transportation routes may have very different views. Their concerns may include the integrity of shipping containers, notification of tribes and states through which shipments are planned, and the capabilities of local emergency personnel to respond to accidents. The U.S. Department of Energy's experience with transporting radioactive wastes has shown that emergency response capabilities are a particular concern in rural and tribal jurisdictions. However, given the small number and size of CAIS items, these transportation issues may not be insurmountable.

People in communities where commercial disposal facilities are located will play a central role in determining the acceptability of this disposal option. According to the Army report, facility owners are not willing to risk arousing negative public reactions for fear of adversely affecting their current business. Because the most widespread concerns are likely to be about health and safety, stakeholders must be convinced that CAIS disposal will not harm the health and welfare of the community (Walsh et al., 1997; Hunter and Leyden, 1995; Freudenberg, 1994). Other documented concerns include geographic equity,[4] acceptance of wastes not originally included in the permit, local involvement in the decision process,[5] and accountability and relationships with program and facility personnel (Hunter and Leyden, 1995; Bradbury et al., 1994). Research has shown that community stakeholders and program managers and their technical staffs frequently differ in their assessments of the reliability and trustworthiness of the organizations responsible for managing and overseeing hazardous waste facilities. Community stakeholders are also likely to be concerned that the technology functions as planned (Wynne, 1992; Hunter and Leyden, 1995; Bradbury, et al., 1994).

Some communities where incinerators are currently located are presumably more receptive to the incineration of potentially controversial wastes, such as CAIS. However, the importance of meaningful public involvement is highlighted by a recent example of a community in Illinois that was assumed to be receptive, but in the absence of prior consultation refused to accept napalm for disposal in a nearby hazardous waste facility. As this example shows, a community may use the political process to intervene in attempts to modify permits. The chances of expanding the current operations of a facility to include CAIS disposal are likely to be improved by prior consultation, existing good relationships, and community confidence in current facility performance, as well as by the facility's economic importance to the community.

Interest Groups. Opposition to incineration of primary or residual wastes is the central issue for members of some citizen groups that have been active in both the non-stockpile and chemical stockpile programs. Key issues previously raised by these groups include potentially harmful environmental and health effects from incinerator emissions, the assessment of incinerator emissions on the basis of trial burns rather than real-time monitoring, and the presence of hazardous residues in the ash. The most serious concerns raised in the past have been the negative environmental and health effects of air emissions—in particular, emissions of PCBs, dioxin, and dioxin-like compounds whose long-term effects, singly or in combination, are uncertain or unknown.[6] More recently, the Non-Stockpile Chemical Weapons Citizens Coalition has called into question the Army's previous estimates of the human toxicity of chemical agents and requested a

---

[4]Committee discussions with stakeholders from Pine Bluff, Arkansas and Tooele, Utah..

[5]Committee discussions with members of the Non-Stockpile Chemical Weapons Citizens Coalition; Hunter and Leyden, 1995; Bradbury et al., 1994.

[6]Recent evidence for these concerns includes "Public Health and Chemical Weapons Incineration" by the Kentucky Environmental Foundation (1998) and information supplied to the committee by the Non-Stockpile Citizens' Coalition, including a letter from the Non-Stockpile Citizens' Coalition to Secretary of Defense William Cohen, dated November 13, 1998. Published reports documenting attitudes toward incineration in the stockpile disposal program include NRC, 1996b, chapter 9; Bradbury et al. 1994; and Smithson, 1994.

reevaluation of "all of the emission limits, exposure limits, and contingency plans for chemical warfare agent."[7] In addition to these substantive issues related to the technology, these groups have long advocated dialogue between the Army and the public and emphasized the need for local involvement in selecting the preferred, site-specific approach.[8]

Members of these groups are committed to using technologies and disposal systems in which all effluent streams are held and tested prior to release.[9] They consider these methods and systems to be less risky than incineration to human health and the environment. In general, these groups do not appear to be opposed to the commercial disposal option *per se*, provided the company has a sound reputation and safety record and that provisions for accountability and acceptance by the local community are in place. However, the number of nonincineration facilities for primary treatment of CAIS may be limited (only one of the five facilities in the Army study).

Even if commercial neutralization is available for primary treatment, incineration is likely to be the facility operator's treatment of choice for residual wastes. In the opinion of one stakeholder group, the Non-Stockpile Citizens Coalition, residual wastes should instead be stored until an acceptable nonincineration technology is developed. During a recent public comment period, this group expressed its strong opposition to the incineration of secondary wastes from Army disposal systems. Groups that share this attitude toward CAIS disposal options are likely to take steps to prevent setting a precedent for incineration, based on their belief that delaying the disposal of CAIS or of residual CAIS wastes does not pose high risks and that it is better to wait for the development of a more acceptable technology than to choose incineration by default.[10] From their perspective, the development of nonincineration technologies by the ACWA program holds out the hope that nonincineration technologies can be used for both the stockpile and non-stockpile programs, as well as for the disposal of other hazardous wastes nationwide.

Commercial Disposal Facilities. The key issue for commercial facility owners is the impact of CAIS disposal on the bottom line. As highlighted in the Army's report, the potential for negative reactions, either from the local community or from outside activists, that could affect the company's current operations is a critical factor. Other concerns include the cost of permit, operational, or monitoring modifications that could outweigh the benefits of CAIS disposal.

---

[7] Comments submitted in February 1999 to the Utah Department of Environmental Quality concerning the RCRA Research Development and Demonstration Permit for the Munitions Management Device (MMD) Version 1; Letter from the Non-Stockpile Chemical Weapons Citizens Coalition to Secretary of Defense William Cohen, November 13, 1998.

[8] Information supplied to the Committee by the Non-Stockpile Chemical Weapons Citizens Coalition. See also Bradbury et al., 1994.

[9] As noted on pp. 159–160 of NRC, 1996a, none of the alternative technologies evaluated by that committee (nor the baseline incineration system for the Chemical Stockpile Disposal System) is "closed loop" in the technical sense that the material is completely recycled internal to the system. The authoring committee reported that its discussions with members of the public indicated that the public used the term "closed loop" in two ways: (1) as a process with few emissions, or fewer unknowns in the emissions; or (2) as a process that allowed all emissions and effluents to be held and tested before being released to the environment. That committee introduced the terminology "hold and test prior to release."

[10] Letter from the Non-Stockpile Chemical Weapons Citizens Coalition to Secretary of Defense William Cohen, November 13, 1998.

Stakeholders Influence on Policy

Stakeholders may be able to influence policy through regulatory and political processes, as well as through the combined impact of these processes and public controversy on the willingness of commercial facility owners to dispose of CAIS. According to current procedures, the public would have three opportunities for review and comment. First, CAIS are classified as lethal chemical agents that are subject to 50 USC 1512 and, therefore, subject to stringent requirements for transportation, disposal, and handling. Attempts to change existing laws and regulations to reduce the complexity and cost of commercial disposal (as suggested by Amr et al., 1998) would open up many occasions for public review. Second, stakeholders would have the right to review and comment on NEPA documents (e.g., environmental impact statements) related to using commercial facilities. Third, stakeholders could comment on changes that may be required to existing state permits to allow disposal of CAIS. Stakeholders opposed to the commercial option could affect policy implementation by legal action or by requesting permit conditions that might make the option more costly and less attractive to a commercial facility.

Politically, local and tribal stakeholders and regional and national interest groups may also have an impact on the final decision through their influence on decision makers in Congress, state legislatures, and federal and state regulatory agencies. At the same time, some local communities and state and federal regulators may be opposed to the long-term storage of residual wastes. Thus, the impact on policy would depend on the relative political strengths of the stakeholders and their representatives.

Issues raised in public discussions, accompanied by possible media coverage, could affect commercial facility owners' willingness to expand into a controversial business area. The combination of public controversy and the increase in costs if changes in permitting conditions are required would almost certainly discourage some facilities from accepting CAIS for disposal.

PROGRAMMATIC CONSIDERATIONS

The Army's report to Congress did not include a discussion of programmatic issues. Examples of issues included in this topic are noted at the end of Chapter 3. The following discussion presents the committee's preliminary views on some programmatic issues.

Schedule

The schedule for the destruction of CAIS material is not expected to be finalized until the best method, or methods, of destruction have been selected. If the RRS option is selected, it is estimated that the program for disposing of stored and recovered CAIS will be completed by the end of the third quarter of fiscal year 2002. However, the chances are good that more CAIS items will be found in the future. Although these could be promptly disposed of by commercial incineration, regulatory hurdles and public resistance could lead to significant schedule delays or disruptions. Thus, although all known CAIS materials could be disposed of fairly quickly, additional materials are likely to be found, making development of a definitive schedule for RRS deployment difficult.

Funding

A "mock contract" developed by the NSCMP has been sent to selected commercial hazardous waste disposal facilities to determine their interest in CAIS disposal. In this mock contract, the Army has estimated a minimum cost of $10,000 per CAIS shipment and a maximum expenditure of $1 million to $4 million per year. Recovered CAIS could be stored temporarily to minimize the total number of shipments. The cost of $10,000 per shipment was based on interviews with these firms. However, it was unclear whether the cost of liability insurance was included in these estimates.

Organizations

The commercial incineration of CAIS will require the concurrence of many organizations who will be both directly and indirectly involved in programmatic decisions that could affect CAIS destruction, schedule, and costs. These organizations include the elected officials at the federal, state, and local levels; base commanders; Army organizations involved in CAIS recovery and transport; and state and federal regulatory agencies. The viability of this option will depend on the effectiveness of the Army in addressing the concerns raised by these organizations.

One of the most significant parties will be the commercial firm. Given the relatively small amount of business and the potential for public controversy associated with the disposal of any chemical warfare materiel, commercial firms may not find CAIS disposal to be an attractive business option.

5

Alternatives to Commercial Incineration of CAIS

In this chapter, the committee evaluates several disposal alternatives to commercial incineration. These include the mobile RRS, which is the Army's "baseline" approach to CAIS disposal; the RRS operating from one or more fixed sites (fixed-mode RRS); and nonincineration technologies at commercial or Army facilities. These alternatives are evaluated in terms of the issues enumerated in Chapter 3.

BASELINE, MOBILE Rapid response system

The committee's evaluation of the mobile RRS alternative is summarized in Table 5-1 and discussed below.

Technology

In principle, the RRS should be a safe and effective method for disposing of recovered CAIS chemicals. The neutralization chemistry on which the RRS design is based has been demonstrated in laboratory studies (U.S. Army, 1997c). However, some questions associated with use of the RRS can only be answered through practical demonstration, and the RRS unit is still undergoing testing in final, integrated form.

Effectiveness and Reliability

The committee observed during an RRS demonstration that the major operations have a laudable simplicity. Most CAIS processing activities are performed manually, and the steps appear to be easily learned by operators and easily controlled. These characteristics contribute to the general reliability of the system.

Several neutralization technologies for destroying chemical warfare agents have been demonstrated as part of the Army's ATA (Alternative Technologies and Approaches) Program and disposal programs by European nations (Shaw and Cullinane, 1998; Yang, 1995). In general, these neutralization processes have proven to be simple, safe, and effective.

The specific chemistry used in the RRS neutralization reactor is closely related to the chemistry used to decontaminate military personnel and equipment under battlefield conditions (Yang et al., 1992). The RRS chemical reagent, 1,3-dichloro-5, 5-dimethylhydantoin, oxidizes sulfur mustard and lewisite to form products that are much

TABLE 5-1 Summary Evaluation of the Mobile RRS Option

	Committee Evaluation
Technology	
Process reliability and effectiveness	Neutralization process is proven; reliability and effectiveness appear to be high; some issues remain unresolved.
Technical maturity	Process chemistry is mature; RRS system is being tested.
Monitoring and disposal of process effluents	Liquid process wastes must be packaged, transported, and treated; liquid wastes must be characterized to ensure safe disposal.
Laws and Regulations	
Consistency with present laws, regulations, and treaties	State-by-state and site-specific RCRA permitting could lead to significant delays and costs.
Costs	
Permitting	Site-specific permit required for each state in which RRS is used; RCRA permit required to store CAIS for more than 90 days.
Indemnification	None
Facility modifications	None
Transportation	Transportability of RRS is a major advantage, but transporting and staffing costs are considerable; treatment of liquid wastes at commercial facilities adds to cost.
Processing operations	Estimated costs of processing (site preparation, set-up, operations, closure) are high; large staff and overhead required; support costs of RRS between deployments required.
Indirect costs	Cost recovery for design and construction; usage fees.
Environmental Impacts, Worker/Public Safety, and Risks	
Environmental impact	Will be assessed during RRS test program and initial permitting.
Worker safety	Will be assessed during RRS test program and initial permitting.
Public safety	Will be assessed during RRS test program and initial permitting.
Risk analysis[a]	Essentially covered in design and development of procedures, costs, etc.; risks of disposition of neutralized wastes unknown but less of a concern.
Public/Stakeholder Involvement	A mobile facility is likely to be more acceptable than a permanent, fixed facility; however, incineration of RRS wastes is strongly opposed by some segments of the public.
Programmatic Aspects	
Schedule	Movement and permitting of RRS could cause delays
Funding	Operational funding requirements are significant.
Organizations	Movement of RRS would require coordination.

[a] Risk analysis includes identifying hazards, understanding the risks, identifying risk control measures, and putting risks into context. The initial discovery of CAIS items, particularly by untrained members of the public, seems to pose the greatest risks. However, the committee's analysis begins at the point of CAIS recovery.

less toxic than the agents, although they are not innocuous. This reagent appears to have been chosen for the RRS because it reacts slowly with the chloroform solvent in some CAIS ampoules but destroys the agents extremely rapidly at low temperatures (25 to 100°C). At these low temperatures, the reactor can operate under pressures only slightly higher than atmospheric pressure (U.S. Army, 1997b). The combination of low reactor pressure and containment of the entire reactor within a controlled-atmosphere enclosure (a glove box vented through a carbon filter stack) minimizes the risk of leaks of agent vapor into the workplace.

The reaction in the RRS reactor is intended to reduce the concentration of agent in solution to less than 50 ppm (i.e., a DRE of 99.9 percent for a 5 percent solution of mustard or lewisite). The products of the mustard reaction include chlorinated sulfoxides, sulfones, chlorinated ethanes and butanes, aldehydes, monochlorodimethylhydantoin, and dimethylhydantoin, all of which are dissolved in the chloroform/tert-butyl alcohol mixture used in the neutralization reaction (U.S. Army, 1997c). The solution of reaction products is drained into a waste drum approved by the U.S. Department of Transportation for storage until it can be transported to a commercial hazardous waste disposal facility for further treatment (to reduce the agent concentration to less than 0.01 percent) and final disposal (U.S. Army, 1998b). An analytical system based on gas chromatography-mass spectrometry analysis has been developed to measure unreacted agents (mustard or lewisite) in the neutralization products. The analysis appears to be sensitive down to approximately 10 ppm of agent. However, there may be some problems in analyzing mixtures of agents (Lucas, 1997).

The composition of the neutralization product solution raises a number of issues that the committee was unable to resolve with the information available:

- Is a reaction product containing up to 50 ppm of mustard or lewisite suitable for transport without further treatment? In the Army's ATA Program, the release standard for mustard in the neutralization effluent is 0.2 ppm, which corresponds to a DRE of 99.9995 percent.
- Can the liquid waste stream be made compatible with nonincineration technologies for secondary treatment of these wastes? For example, can glass fragments and other solids be filtered from the neutralization mixture and readily treated?
- If items in a CAIS set are broken or leaking and the packing material has been contaminated, how would this material be processed?
- Does the working environment of the RRS glove box allow for an effluent sample to be held until it is analyzed and for the effluent to be reprocessed if it does not meet the release standard?
- Are special toxicology issues associated with some reaction products, such as bis(2-chloroethyl) sulfoxide, that are formed in the mustard neutralization reaction?
- Are special disposal issues associated with the major arsenic-containing product, 2-chlorovinylarsonic acid, produced by the oxidative neutralization of lewisite?
- Are special disposal issues associated with chlorinated hydrocarbon by-products, such as 1,2-dichloroethane, which is a so-called "land-banned chemical" subject to special regulatory restrictions?
- Will the neutralization process deal effectively with the solid deposits (e.g., cyclic sulfonium salts) often found in old samples of sulfur mustard?

Technical Maturity

The RRS concept includes aspects of many established operations in the Army's chemical demilitarization operations. For example, the Army has extensive experience using chemical systems for destroying chemical warfare agents and using monitors to detect their presence in the atmosphere. The Army also has extensive experience with the handling and transportation of toxic materials.

Although the RRS design draws heavily on the Army's experience, the overall system is only now being assembled and tested as an integrated system. Until testing has been completed, a final judgment about many facets of its operation would be premature. For example, problems may be encountered in analyzing the incoming samples in the unpacking and characterization compartments of a working RRS if the atmosphere contains high concentrations of organic vapors. The current assumption that the liquid waste will be treated by incineration may have to be reconsidered. Solving these problems may increase development time and cost significantly. A flow chart, or similar analytical tool, that captures all the possible paths from CAIS feeds to final disposal states should be developed and made available to all parties involved in evaluating the RRS and approving its use.

Monitoring and Disposal of Process Effluents

The efficacy of the monitoring systems used throughout the RRS apparatus can only be evaluated when the system is completed and in operation with actual CAIS materials. Analysis of the neutralization effluents may prove to be a challenge, especially if the release standard is lower than the current 50 ppm. In that case, the methods used for analyzing the water-based effluents from the neutralization process developed in the ATA Program may not be adaptable to the chloroform-based effluents generated in the RRS.

A critical point concerning the disposal of RRS process effluents is that they will require further processing for ultimate disposal. The currently proposed approach is incineration of these effluents in an approved commercial facility.

Laws and Regulations

Use of the baseline, mobile RRS for CAIS disposal is likely to require obtaining a RCRA operating permit for each state, and even for each site within a state. Permit conditions are likely to vary from state to state and perhaps from site to site. Because there may be dozens of CAIS recovery sites, and because many months are often required to obtain RCRA operating permits, this requirement could lead to significant delays in CAIS disposal unless some process can be established to expedite approvals for sites within a state and across states. The transportability of the RRS, which solves a number of difficult issues related to transporting CAIS prior to characterization and treatment and to transferring one site's CAIS problem to another site, is a favorable feature that could help expedite the permitting process.

Costs

Permitting

For every non-Superfund site where CAIS are found and the RRS is used, a RCRA permit will be required. The cost of obtaining a RCRA permit has been estimated to be $250,000 by the Army (U.S. Army, 1997a, p. F-13), and, based on the Army's experience in obtaining a RCRA permit to test the RRS in Utah, can take up to three years. The time required to obtain RCRA and other permits in each state where CAIS are found may mean the use of the mobile RRS is impractical. Also, state-specific permit restrictions may limit the use of the RRS to a single campaign to destroy known CAIS items; a permit modification (or new permit) may be required to process CAIS items found after the permitted RRS operation has been completed. If the simplicity and transportability of the mobile RRS option are appealing to regulators and community stakeholders, the costs in time and resources to obtain individual permits could be reduced by an effective public involvement program.

Transportation

The costs and logistics of transporting the RRS (two trailers and a mobile analytical laboratory), as well as supplies and staff, could limit the use of the RRS as a rapid disposal facility. Transport by air would require two C-141 aircraft. Transport by land would entail trucking the RRS trailers and other equipment. The Army estimates the cost of moving the RRS and associated equipment from Tooele, Utah, to Anchorage, Alaska (2,500 miles) to be $33,000. Land or air transportation would also involve moving the staff required to operate the RRS. These costs were estimated by the Army to be more than $172,000 for 55 calendar days of operation of the RRS in Alaska (U.S. Army, 1997a, p. F-14).

Packaging

Under the baseline RRS approach, the costs of identifying CAIS items, separating them into industrial chemicals and chemical warfare material (sulfur mustard and lewisite), and repackaging them would be borne by the Army, typically through the environmental restoration line item in the budget of the installation where the CAIS are found. For CAIS found on former military bases, these costs will be paid from a special environmental restoration fund reserved for inactive military sites.

Processing

The principal cost with the RRS would be the cost of processing CAIS items. In the cost estimate for the proposed treatment of CAIS at Fort Richardson, Alaska, 13 RRS-specific staff and several on-site Army staff were required for processing operations. The processing of the seven PIGs at Fort Richardson, including mobilization, site preparation, set-up, processing, closure, and site cleanup was estimated to take 55 calendar days and to cost almost $457,000 in labor costs alone (average $64 per hour). The cost of materials and equipment, not including transportation and usage fees, was estimated to be another $228,000. Finally, management, engineering, and other costs were estimated to be more than $250,000.

Based on the one-gallon capacity of the RRS neutralization reactor, the Army assumed that the RRS can process one CAIS bottle or three ampoules per batch and that the time required for neutralization of a batch is 15 minutes (30 minutes for the CAIS types consisting of agent on charcoal). The processing rate is not a cost issue, per se, although the daily labor cost of operating the RRS (more than $13,500 per day in Alaska) is significant.

Indirect Costs

The costs of designing, building, and maintaining the RRS, as well as the costs of replacement and spare parts, were included in the cost estimate of RRS operations and were estimated to be $5,100 per calendar day. This cost, along with a usage fee of $2,150 per day for the mobile analytical laboratory, came to almost $400,000 in materials and equipment costs for disposing of the seven PIGs during a 55-day campaign.

Environmental Impacts, Worker/Public Safety, and Risks

Many potential environmental and safety issues will be evaluated during the RRS test program in Utah, and during the site-specific permitting process for RRS operations. In general, because CAIS disposal via the RRS would be implemented by the Army, the risks of recovery, treatment, and disposal were probably addressed in the planning and design process.

The risks of storage, handling, and treatment are controlled through the secure storage areas, glove boxes, monitoring systems, excess neutralization capacity, and other features of the RRS design. The Army's Technical Escort Unit is assumed to be responsible for the recovery, packaging, and transport of the CAIS. The two areas that have not been covered explicitly in the design are the transport and disposal of the neutralized wastes, although these wastes will be similar to other regularly handled industrial wastes. Because risk reduction, mitigation, and control measures are built into the process design and procedures of the RRS, as well as the proposed staffing plans, risks are accounted for in the cost and schedule estimates.

Public/Stakeholder Involvement

The issues about public acceptability raised in Chapter 4 can also be applied to the options for CAIS disposal discussed in this chapter, including the RRS options.

Key Stakeholders

With the exception of commercial facility owners, the stakeholders for the mobile RRS option are the same as for the commercial incineration option. Stakeholders include the Army and Congress, regulators, local populations near the RRS and along transportation routes (which will differ from those for the commercial option), and national interest groups. Although the basic issues for each stakeholder group are the same as for the commercial disposal option, the particular application of each issue and the potential critical issues vary with the option under consideration.

Key Issues for Each Stakeholder Group

Army and Congress. An issue of primary concern to both Congress and the Army is that the mobile RRS is not cost efficient for treating small quantities of CAIS. However, one advantage of the mobile RRS is that it avoids the need for construction of a permanent disposal facility and the need to transport CAIS. Transportable facilities minimize the impacts on active military bases and may make this option more acceptable to the public than a fixed RRS.

State Legislators and Regulators. The mobile RRS has several features favoring its acceptability to the constituencies represented by legislators and regulators. First, it provides a means of eliminating risks from recovered CAIS to nearby communities without transferring the risks elsewhere (for example, transportation risks en route or disposal risks at a distant facility). Second, it avoids the potential for negative public reaction to permitting changes that could be required for commercial incineration facilities. Third, it eliminates concerns about location of a permanent facility and safe transportation. This option would, however, require storage of CAIS, which could raise concerns among nearby communities.

Local Populations. On balance, the mobile RRS appears to be more positive for communities near recovered CAIS than either commercial disposal or the fixed RRS option. Three major advantages are (1) it does not use incineration; (2) it is not a fixed facility (which might be used for other disposal purposes), and it can be removed quickly following on-site disposal of CAIS; and (3) it avoids the risks of transporting CAIS and the associated handling risks. Issues that could cause concerns among nearby communities are (1) the extended period of storage pending deployment of an RRS; and (2) the storage of residual wastes, if these are stored until a nonincineration technology becomes available.

Regional and National Interest Groups. The mobile RRS meets several of the stated acceptability criteria of regional and national interest groups. It uses neutralization rather than incineration. It is a temporary rather than a permanent facility, and it does not require the transportation of chemical agent off the site where CAIS are discovered or stored. It also enables an affected community to take care of its own waste rather than placing the burden on another community.

However, a major concern of these groups is the proposed use of incineration technology for the treatment of residual wastes. In their view, using incineration would establish a precedent for continued reliance on this technology and would undermine the urgency of developing more acceptable alternatives.

Stakeholder Influence on Policy

Local, regional, and national stakeholders can affect policy through political and regulatory processes. The mobile RRS offers several advantages in terms of regulatory requirements. Transporting the technology to the waste, rather than transporting the waste to the technology, eliminates the need to comply with complex transportation requirements that could arouse public concerns. In addition, it eliminates the need for statutory or regulatory changes that might be necessary for commercial incineration. However, stakeholders may still have opportunities for review and comment on project-specific NEPA documents. They may also intervene in the state permitting processes that would be required for the RRS and for on-site storage of CAIS pending deployment of the RRS to the site. As noted in Chapter 4, comments recently submitted by the Non-Stockpile Chemical Weapons Citizens Coalition to the state of Utah indicate a likelihood of strong opposition to the use of incineration as a secondary technology.

The Army could resolve this problem in two ways. First, the Army could apply for permits for storing CAIS or retaining residual wastes following neutralization on site or shipping them off site to an interim storage facility (either commercial or government-owned) until an acceptable nonincineration technology becomes available. However, this approach entails several potential disadvantages: (1) it is uncertain when an acceptable technology will become available; (2) Congress and the Army would be faced with additional monitoring costs; (3) federal and state regulators may be reluctant to approve long-term monitoring of wastes for which a technical solution (incineration) is already available; and (4) communities near a proposed interim storage facility may oppose the facility (e.g., concerns about health and safety, equity, and becoming a permanent dumping ground).

A second approach would be to engage the leaders of interest groups and the local community with Army representatives in developing "win-win" solutions. This approach has the advantage of bringing together key group members and the Army personnel who have already established working relationships with them. A possible disadvantage of this approach is the uncertainty about the storage time required until an acceptable technology becomes available.

Programmatic Aspects

Movement of the RRS between states, and between sites, will require significant coordination among the Army, federal and state regulators, and other state and local government officials. A lack of coordination could lead to delays and add to the cost of CAIS disposal, and thus add to the funding requirements for the program.

FIXED RAPID RESPONSE SYSTEM

The committee also evaluated the use of the RRS in a fixed mode of operation. (It also briefly investigated modified RRS options, such as the Army's ECS [Expedient CAIS Disposal System], which is described in Box 5-1). To avoid permitting and other site-specific costs associated with moving the RRS to CAIS discovery sites, the committee considered the alternative of having one or more fixed RRSs at permitted storage sites for CAIS materials. In this scenario the CAIS materials would be transported to the nearest RRS (see Table 5-2).

Technology

The technical issues for the fixed RRS alternative are the same as for the baseline, mobile RRS option.

> BOX 5-1 Expedient CAIS Disposal System
>
> Disposal of CAIS via a modified RRS, called the Expedient CAIS Disposal System (ECS), would be an alternative to the RRS in some situations. Although the ECS has substantial technical limitations, using it might result in cost savings in terms of reduced permitting needs because of its rapid deployment to a CAIS recovery site and its reduced staffing and support needs. The usefulness of the ECS would be limited by its small glove box, its inability to remove CAIS items from metal overpacks, its inability to remove neutralization wastes from the reactor under engineering controls, and other factors. Nevertheless, the ECS could be applicable in some situations.

Laws and Regulations

Operation of the RRS in a fixed mode offers significant regulatory advantages over the mobile RRS option. Once an RRS is sited and granted a long-term operating permit at one or perhaps a few regional sites, no additional state-by-state or site-by-site operating permits would be necessary, which would be a significant cost and schedule advantage over the mobile RRS. However, use of the RRS in a fixed mode would require transporting recovered CAIS, as found, to the RRS. This transportation would require transport plans approved by the U.S. Department of Health and Human Services and by state governors.

Costs

Most of the costs for the fixed RRS are the same as for the baseline, mobile RRS. The key differences in cost are for permitting; packaging and transporting the CAIS sets and items; processing; and recovery of indirect costs. Each of these factors is discussed below.

Permitting

RCRA and other permits would only be necessary for the relatively few states in which fixed RRSs would be located, which would result in substantial savings in permitting costs and time. The savings are based on the assumption that the permits would allow out-of-state CAIS items to be destroyed in the RRS and would allow the RRS to process CAIS items that have not yet been recovered and stored in the state. Neither of these assumptions would be valid in Utah, the only state that has issued a RCRA permit for RRS operations so far. State permit requirements vary, however, so these factors may not necessarily make the fixed RRS a less attractive option in terms of cost. They should be kept in mind, however, when considering this alternative.

Transportation

Although the CAIS materials would have to be brought to the fixed RRS, transportation costs are modest. In its cost analysis, the Army estimated that the cost of

TABLE 5-2 Summary Evaluation of the Fixed RRS Option

	Committee Evaluation
Technology	
Process reliability and effectiveness	Neutralization process is proven; reliability and effectiveness appear to be high; some issues remain unresolved.
Technical maturity	Process chemistry is mature; RRS system is being tested.
Monitoring and disposal of process effluents	Liquid process wastes must be packaged, transported and treated; liquid wastes must be characterized to ensure safe disposal.
Laws and Regulations	
Consistency with present laws, regulations, and treaties	RRS permitting requirements by states and EPA are reduced; approvals for CAIS transportation by U.S. Department of Health and Human Services are increased.
Costs	
Permitting	Several operating permits are necessary for RRSs; permits may limit use to in-state or known CAIS items; permits and transportation plans are required to ship CAIS.
Indemnification	None
Facility modifications	None
Transportation	Transporting CAIS to RRS with escorts is an added cost, but field staffing costs are lower; treatment of liquid wastes at commercial facilities adds cost.
Processing operations	No site preparation or closure costs; in-field costs of characterizing, separating, and packaging CAIS would be incurred.
Indirect costs	Cost recovery for RRS design and construction; usage fees.
Environmental Impacts, Worker/Public Safety, and Risks	
Environmental impact	Will be assessed during RRS test program and initial permitting.
Worker safety	Will be assessed during RRS test program and initial permitting.
Public safety	Will be assessed during RRS test program and initial permitting; transportation to fixed RRS must also be assessed.
Risk analysis[a]	Essentially covered in design and development of procedures, costs, etc.; risks of disposition of neutralized wastes unknown but less of a concern.
Public/Stakeholder Involvement	Incineration of RRS effluents is strongly opposed by some segments of the public; Army should seek public approval of RRS sites.
Programmatic Aspects	
Schedule	CAIS transportation approvals may cause limited delays.
Funding	Operational funding required.
Organizations	RRS sites would have to be approved by base commanders.

[a]Risk analysis includes identifying hazards, understanding the risks, identifying risk control measures, and putting risks into context. The initial discovery of CAIS items, particularly by untrained members of the public, seems to pose the greatest risks. However, the committee's analysis begins at the point of CAIS recovery.

transporting seven PIG overpacks by military aircraft from Fort Richardson, Alaska, to the permitted storage site at DCD in Utah and the cost of temporary storage to be $33,300. The same cost for moving the PIGs to the permitted storage site at Pine Bluff, Arkansas, was estimated to be $45,300. The associated staff travel and per diem costs were about half of the costs of transporting the RRS to the CAIS location in Alaska. Additional costs could be incurred, however, in moving chemical warfare materiel rather than moving the processing equipment, especially if transportation plans must be prepared and regulatory agency approvals obtained.

Processing and Indirect Costs

The same issues and concerns that were raised for the mobile RRS apply to the fixed RRS because the CAIS handling and processing operations would be the same.

Environmental Impacts, Worker/Public Safety, and Risks

Use of the RRS in a fixed mode would require that all recovered CAIS be transported from recovery sites to the RRS, as recovered, which would increase risks to the public. However, using the Army's Technical Escort Unit to transport recovered CAIS could minimize transportation risks.

Public/Stakeholder Involvement

Key Stakeholders

The stakeholders for the fixed RRS are the same as for the mobile RRS, although the local populations would differ. In this option, local populations would be located near a fixed RRS, near an interim storage facility, and along proposed transportation routes.

Key Issues for Each Stakeholder Group

The basic issues for each stakeholder group are the same as for commercial incineration or the mobile RRS, although the particular applications and relative importance of issues may vary.

Army and Congress. The fixed RRS offers a cost advantage over the mobile RRS by processing CAIS at fewer locations. Compared with commercial incineration, it avoids potential negative public reaction and controversy over attempts to change the law or regulations. However, the cost savings may be less than anticipated because siting a permanent facility might require more funding and public involvement than the mobile RRS would. Three issues are likely to arise for the fixed RRS: (1) concerns of local communities about the health and safety impacts; (2) concerns about the economic impacts of a permanent, fixed facility; and (3) concerns about transporting CAIS (this concern may be reduced once the small numbers and amounts of CAIS are made known). An additional issue for the Army is that the selection of RRS location(s) may have to focus on sites at existing chemical weapons disposal facilities; thus coordination with the Chemical Stockpile Disposal Program will be essential.

State Legislators and Regulators. The fixed RRS option would address public concerns and political pressures to remove CAIS risks from communities near the sites where CAIS are recovered but not processed. However, this option is likely to arouse new concerns in communities along CAIS transportation routes, near interim storage facilities, and near the fixed-RRS sites.

National and Regional Interest Groups. Although the fixed RRS would use neutralization to treat CAIS materials, incineration is the most likely technology for treatment of residual wastes. Incineration has aroused much public opposition among national and regional environmental groups. In addition, a key issue for the Army will be balancing short-term and long-term program goals and nurturing the effective relationships that have begun to develop with the stakeholders. Interest groups may also be concerned about the risks of CAIS transportation.

Local and Tribal Populations. Although communities near CAIS discovery sites will appreciate the removal of the risk of untreated CAIS, the transfer of risk to other communities is likely to be a concern. Communities along transportation routes may be concerned about the risks of transporting CAIS and the capabilities of local and tribal emergency responders (again, this opposition may be limited because of the small amounts of CAIS involved). Communities near proposed RRS site(s) may raise issues related to the siting of RRSs, including equity (more than one RRS), health and safety, involvement of local communities in decision making, fears that the RRS will become a site for treatment of other wastes, and Army accountability.

Stakeholders Influence on Policy

Stakeholders will have several opportunities to influence policy: during the selection of site(s) for the fixed RRSs, during the application process for the operating permit for the RRS, and during regulatory oversight of the transportation of CAIS to the RRS. U.S. Department of Health and Human Services requirements for notification prior to the shipment of CAIS and U.S. Department of Transportation regulations that require appropriate placards would increase the public visibility of shipments.

Programmatic Considerations

The fixed RRS would have some programmatic impacts. First, the Army would have to work with military base commanders, regulators, and affected stakeholders to select the site(s) for the RRS(s). Permits for operation at the site(s) would have to be obtained. Transportation of as-recovered CAIS would require coordination of transportation resources (i.e., Technical Escort Unit personnel and equipment) and funding. Close coordination between the Army, regulators, and the affected state and local governments would be necessary.

NONINCINERATION ALTERNATIVES

The committee's evaluation of nonincineration alternatives for CAIS disposal is summarized in Table 5-3 and discussed below.

TABLE 5-3 Summary Evaluation of Selected Nonincineration Options[a]

	Committee Evaluation
Technology	
Process reliability and effectiveness	Neutralization proven during stockpile program development; other processes are under development or unproven.
Technical maturity of the process	Some commercial processes exist; agent-specific treatment processes are under development.
Monitoring and disposal of process effluents	Unknown, but, no monitoring is expected beyond routine analysis of residual wastes prior to release; addition of agent monitors could be an added cost at a commercial facility.
Laws and Regulations	
Consistency with present laws, regulations, and treaties	Requires legal/regulatory relief, clarification, or flexibility; some facility permit modifications may be required.
Costs	
Permitting	Some permits required for all CAIS disposal alternatives.
Indemnification	A potential added cost to the Army or facility.
Facility modifications	Monitoring and other modifications may add cost.
Transportation	Transportation to commercial sites may add cost; military escorts may be required; not clear how handling, characterization, and transportation costs are funded.
Processing operations	CAIS packaging would add costs; dedicated processing of CAIS could be costly.
Indirect costs	Hidden indirect costs (overhead, administration, maintenance).
Environmental Impacts, Worker/Public Safety, and Risks	
Environmental impact	Air/water emissions minimized by nature of process.
Worker safety	Will be assessed after technology identified and tested.
Public safety	Will be assessed after technology identified and tested.
Risk analysis	Potential risks should be considered in the specification of any treatment option, especially for storage and handling of CAIS items.
Public/Stakeholder Involvement	Nonincineration-based methods likely to be more acceptable to many members of the public.
Programmatic Aspects	
Schedule	Significant delays possible during technology development or identification; more rapid disposal schedule possible once available.
Funding	Significant funds required for any technology development program.
Organizations	Corporate commitment is unknown.

[a] This summary assumes that lewisite could be treated by the technology already in use in the Army's CAMDS (Utah) facility for destruction of bulk lewisite (the Canadian Swiftsure process—neutralization followed by immobilization of the arsenic-containing products in a cement-like matrix that is subsequently disposed of in a landfill). Sulfur mustard could be treated by the technology to be used in the chemical stockpile disposal facility being built at Aberdeen Proving Ground (Maryland) for destruction of sulfur mustard in ton containers (neutralization, followed by biodegradation), or technologies at other commercial facilities could potentially be used.

Technology

Both the Chemical Stockpile Disposal Program and the ATA Program have demonstrated technologies for the disposal of sulfur mustard and lewisite that do not involve incineration of either CAIS chemicals or effluents from initial treatments. Additional technologies are being tested in the ACWA (Assembled Chemical Weapons Assessment) Program. Some of these processes may be directly applicable to CAIS materiel. The Army might consider using facilities of the Chemical Stockpile Disposal Program (if permitted by changes in statutes), commercial facilities, or small, government-owned facilities dedicated to CAIS disposal. Treatment in a chemical surety laboratory might be feasible for isolated finds of CAIS ampoules. Existing procedures for laboratory-scale disposal of agent residues with oxidative reagents, such as bleach or persulfate, should be applicable to small quantities (up to seven grams) of blister agents, either neat or in solution. Laboratory disposal might be acceptable if it were limited to finds of just a few ampoules of solution (certainly less than a full or nearly full CAIS), if there were an established route for disposal of CAIS now in storage or for finds of full or nearly full CAIS. Laboratory disposal thus represents a supplementary approach for removing a hazard quickly and efficiently, but it is not a comprehensive solution to the problem of CAIS disposal. Mustard or lewisite adsorbed on charcoal may require different treatment than neat chemicals or solutions.

Process Reliability and Effectiveness

Both sulfur mustard and lewisite have been neutralized successfully with water or aqueous alkali. In fact, neutralization with hot water will be used to destroy 1,625 tons of "stockpile" HD stored at Aberdeen Proving Ground in Maryland. The reaction with water produces an aqueous solution of thiodiglycol, a relatively innocuous commercial chemical. The mustard concentration is reduced to less than 0.2 ppm, which corresponds to a 99.9995 percent DRE. However, the thiodiglycol solution must undergo further treatment because it is defined as a Schedule 2 precursor compound under the CWC, which means it must be monitored until it is destroyed to ensure that it is not reconverted to sulfur mustard. The thiodiglycol will be destroyed in a biological reactor closely analogous to a sewage treatment plant (NRC, 1996a). The effluent from the bioreactor has very low toxicity to mammals and aquatic species and meets CWC and Maryland standards for destruction and disposal. After treatment in a federally owned treatment works, the effluent will be released into Chesapeake Bay.

Ten "ton containers" of lewisite stored at the DCD are scheduled to be destroyed by neutralization in a facility recently built and permitted at the Chemical Agent Munitions Disposal System (CAMDS), a pilot-scale facility located at DCD.[1] The CAMDS process involves oxidation of the agent by hydrogen peroxide followed by neutralization with aqueous alkali (Maggio, 1998). Based on Canadian experience with this technology, the agent concentration in the effluent from treating neat agent should be reduced to less than 0.09 mg/ml. The effluent will be prepared for disposal by immobilization in a cement-silica grout. The process appears to meet CWC and Utah standards for destruction and disposal.

Neutralization is generally a reliable, robust technology for agent destruction. The Aberdeen and CAMDS processes both operate at near- atmospheric pressure and at temperatures below 100°C. Such mild conditions greatly reduce the danger of dispersing

---

[1] Utah State Department of Environmental Quality. 1998. Permit I. D. Number 5210090002. April 16.

agent vapor in the event of loss of containment (e.g., a reactor leak). The same conditions facilitate safe shutdown of the reactor in the event of an electrical failure or stirring problems. With neutralization, the reactor contents can be retained until agent destruction is confirmed.

The processes of unpacking, draining, and decontaminating containers are more complex for a neutralization process than for incineration. Metal containers must be cut, punched, or disassembled, and glass containers must be crushed in a glove box to contain agent vapors. Decontamination of metal and glass residues, as well as packing materials (e.g., sawdust) can be done most easily by burning or thermal treatment although nonthermal decontamination is planned for the ATA Program.

Technical Maturity

Neutralization of sulfur mustard has been tested on a significant scale (114 liter reactor) in the ATA Program (NRC, 1996a) and will be pilot tested on a "production scale" in the Aberdeen facility. This neutralization technology involves detoxification of the agent with hot water. An alternative approach to neutralization of sulfur mustard is based on treatment with monoethanolamine (MEA) or with glycol mixtures (Petrov et al., 1998). The MEA procedure has been extensively tested at laboratory scale in Russia and will be pilot tested at a facility being built at Gorny in the Saratov region (Kovalyev, 1997). Gas-phase hydrogen reduction, as discussed in the Mitretek report (Amr et al., 1998), has been successfully demonstrated with sulfur mustard on a scale of 780–870 g., roughly equivalent to processing seven 4 oz. ampoules or bottles of HD (Kummling et al., 1999).

In Canada's Project Swiftsure in 1990–1991, 2.5 metric tons of lewisite were treated by oxidation followed by hydrolysis (McAndless et al., 1992). Hydrolysis of lewisite followed by electrochemical reduction of the arsenic-containing wastes has been selected for pilot testing in Russia (Kovalyev, 1997). Destruction of lewisite by gas-phase hydrogen reduction was demonstrated in Russia but was found to be "very unsafe for the working staff and detrimental to the environment" (Petrov et al., 1998).

Monitoring and Disposal of Process Effluents

Neutralization processes produce mostly liquid wastes, plus varying amounts of solid waste in the form of decontaminated packaging materials. The gaseous effluents from neutralization are usually small and are handled by venting through charcoal filters. The metal or glass container materials can be decontaminated by thorough washing with hot water, dilute caustic, and/or bleach solution. Prior to disposal, they can be monitored for the presence of residual agent by holding them in a closed chamber with an ACAMS monitor, as is done for empty ton containers in the ATA Program.

Liquid waste streams from neutralization generally undergo further treatment before final disposal. In the ATA Program, the effluent from hydrolysis of sulfur mustard is analyzed to confirm a satisfactory level of agent before the liquid is released for further processing and disposal (NRC, 1996a). It remains to be seen whether such a monitoring system would be necessary in a commercial neutralization facility. With arsenic-containing agents, such as lewisite, it is desirable that the arsenic content be immobilized as an insoluble arsenate salt (such as ferric arsenate) before disposal in a landfill. In the CAMDS process, the toxic arsenate-containing waste stream from lewisite neutralization will be immobilized in a solid grout before final disposal (Maggio, 1998).

Specific Applications of Nonincineration Processes

Sulfur Mustard. At the Aberdeen facility, the treatment of sulfur mustard, either pure or in solution, would entail the following steps:

- hydrolysis by vigorous stirring with hot (90°C) water to generate a dilute solution of thiodiglycol and hydrochloric acid
- adjustment of the solution acidity to near-neutral pH with aqueous sodium hydroxide
- steam stripping or air stripping to remove volatile organic impurities (mostly chlorinated hydrocarbons)
- biological oxidation of thiodiglycol and other organic components of the aqueous solution
- discharge of the aqueous effluent into the sewage treatment plant at Aberdeen Proving Ground for final cleanup before disposal

It seems likely that a chloroform solution of mustard could be treated similarly. The chloroform would distill from the neutralization effluent along with other volatile chlorinated hydrocarbons that are normally present.

Monitoring and waste disposal for CAIS might be the same as in the Aberdeen process. The major modifications in procedure would relate to unpacking CAIS items and decontaminating the glass and metal residues. The latter might be carried through the ton container clean-out line in a basket, as is done in the Aberdeen process for small metal parts, such as valves and fittings.

The following points would have to be verified for the destruction of CAIS mustard:

- establish that chloroform can be treated like other chlorinated hydrocarbons in the Aberdeen process
- verify the efficacy of the biodegradation process because the sulfur mustard in CAIS items may have different impurities than those found in ton containers stored at Aberdeen
- demonstrate the effectiveness of the current ACAMS and Depot Area Air Monitoring System (DAAMS) air monitors in the modified process

Lewisite Solutions. Lewisite solutions should be amenable to destruction by the CAMDS process. The process would involve the following major steps (Maggio, 1998):

- oxidation of lewisite to chlorovinylarsonic acid with aqueous hydrogen peroxide
- catalytic decomposition of excess hydrogen peroxide
- hydrolysis of the chlorovinylarsonic acid with hot aqueous alkali to form sodium arsenate and acetylene (gas)
- analysis of the neutralization product to show that the agent concentration is below 1 ppm

The neutralization product, an aqueous solution of sodium chloride and sodium arsenate, would be shipped off site for immobilization in a cement-silica grout, which would then be deposited in a hazardous waste landfill. It seems likely that a chloroform solution of lewisite could be processed similarly. The chloroform, which should distill before the treatment with hot alkali, could be condensed in the vent gas knockout drum or collected on the vent filter for commercial disposal. In addition to the questions listed above for the neutralization of mustard, the following questions unique to lewisite processing would have to be resolved:

- effectiveness of the air monitoring systems for lewisite (ACAMS and DAAMS)
- requirements for disposing of the arsenate-containing effluent

Blister Agents on Charcoal. Bottles containing charcoal on which lewisite or sulfur-mustard is adsorbed present a special problem. The water-based Aberdeen and CAMDS processes may not be effective for removing the agent completely from the carbon matrix, which is likely to be poorly saturated by the aqueous reagents. (The neutralization process in the RRS uses an organic solvent to overcome this problem.) If the aqueous processes do not prove to be effective with agent-on-charcoal samples, other approaches are available, especially because the CAIS items containing only solids pose relatively low risks. If these materials were reclassified simply as hazardous waste rather than as lethal chemical weapons, they could be transported to commercial TSDFs for incineration with very little risk to the public, the workforce, or the environment. Otherwise, the agent-on-charcoal CAIS could be stored safely until appropriate disposal processes were proven. Some disposal options are listed below.

Gas-phase hydrogen reduction. One company in the Army's survey uses high-temperature, gas-phase hydrogen reduction in facilities outside the United States (Amr et al., 1998). In testing under the ATA Program, the process destroyed sulfur mustard effectively on a laboratory scale. A similar Russian-developed process also destroys lewisite but appears to be problematic because of questions about the fate of arsenic in the reaction effluent (Petrov et al., 1998). The arsenic might exit the reactor as metallic arsenic or arsine gas, both of which are toxic. Extensive research and development may be required to establish an effective means of removing these materials from the effluent. Neither process has been reported to destroy agent adsorbed on charcoal, but they might work because hydrogen readily penetrates porous solids. The gas-phase hydrogen reduction process would require extensive development to demonstrate its effectiveness. One open issue is whether the hydrogenation reactor should be placed in an enclosure, which would contain agent vapors but would introduce the new risk of accumulating hydrogen (from leaks). High concentrations of hydrogen could be an explosion hazard (NRC, 1996a).

Two-stage Russian chemical agent disposal process. In the first stage of this process, the reaction of MEA or MEA-glycol mixtures with sulfur mustard cleaves the carbon-chloride (C-Cl) bonds in the agent molecule that are associated with its toxicity (Petrov et al., 1998). Similar reactions with lewisite break the arsenic-chlorine (As-Cl) bonds associated with vesicant activity although they do not affect the immutable toxicity of arsenic. The MEA treatment is also said to be effective for mixtures of mustard and lewisite (Kovalyev, 1997). In the second stage of the disposal process, the viscous reaction mass from the MEA treatment is heated with bitumen. Moderately high temperatures (ca. 200°C) and vacuum are used to distill excess MEA for recycling. After cooling, the bituminous mixture forms a hard black solid that is said to be suitable for landfill disposal. This two-stage process may be effective for deactivating agent-on-charcoal and providing a matrix for landfill disposal. Testing would be necessary to establish that toxic materials do not leach from the bituminous product.

SCWO (supercritical water oxidation). Oxygen dissolved in water at 400 to 600°C under high pressure is a powerful oxidant that destroys most types of organic chemicals. SCWO technology is being tested by the Navy for disposal of shipboard wastes and is being demonstrated by the ATA Program for final treatment of the neutralization product of VX nerve agent. Another neutralization and SCWO facility is being funded by the ACWA Program (NRC, 1998). The Army has also contracted for a pilot-scale SCWO unit to be installed at the Pine Bluff Arsenal for the disposal of wastes from its smoke and obscurants program. SCWO has been demonstrated to destroy sulfur mustard with a 99.9999 percent DRE at laboratory scale (Spritzer et al., 1995).

The Army has proposed that agent-containing charcoal filters from stockpile disposal facilities, such as Aberdeen Proving Ground, be disposed of by SCWO. The filter materials would be pulverized and fed as an aqueous slurry to a SCWO unit. The expectation is that this treatment would destroy both the charcoal and the adsorbed agent, producing carbon dioxide and an aqueous solution of inorganic salts. The latter could be retained to confirm complete destruction of the chemical agents. If this approach is successful, it would provide an attractive nonincineration option for the disposal of CAIS containing agent-on-charcoal material. For lewisite-on-charcoal, the aqueous arsenate product could be immobilized for disposal, as it is in the CAMDS process.

Laws and Regulations

The use of commercial facilities for CAIS disposal will require changes, clarifications, or more flexibility in existing laws and regulations. Current regulations and legal interpretations mandate that the Army maintain control of CAIS materials for transportation and disposal. Even if the transport and disposal of CAIS could be accomplished without Army facilities or personnel, and assuming that a commercial facility using nonincineration technology were available and had obtained, or could obtain, an operating permit, regulators might require modifications to the facility's operating permit. Army nonincineration-based facilities might also require changes to their operating permits.

Use of the Army's stockpile disposal facilities for CAIS disposal, even those facilities employing nonincineration technologies, would require Congressional and executive action to modify the existing legal restrictions that prohibit their use for disposal of any non-stockpile materiel. Public acceptance of a plan for disposing of CAIS at stockpile facilities is likely to be a prerequisite for congressional action.

Costs

Permits and approvals will have to be obtained for any nonincineration disposal method. For new technologies, research and development permits may have to be obtained, and testing may be required prior to permitted operations. All nonincineration disposal options involve either bringing a portable disposal facility to CAIS or bringing the CAIS to a fixed disposal facility. In either case, costs similar to those for CAIS and RRS transport will be incurred. Also, the characterization, separation, and repackaging of CAIS items will be required unless all CAIS items can be processed together in the nonincineration facility.

For CAIS found in metal overpacks, methods of accessing the CAIS items inside will be required, and estimates for these costs must be included in overall cost estimates. Other costs would be incurred for modifications to a facility already in operation if it is not equipped to receive and process CAIS. Costs for facility modification and personnel training for new facilities and equipment designed or easily modified to process CAIS may be minimal.

Finally, disposal costs will vary with the alternative. The cost estimates for each CAIS disposal alternative must include indirect costs, such as management, overhead, depreciation, and maintenance.

Environmental Impacts, Worker/Public Safety, and Risks

The impacts of nonincineration-based disposal technologies on risks to the environment and the safety and health of workers and the public will depend on the particular technology and are difficult to predict for a general case. However, air emissions would generally tend to be less for nonincineration technologies than for incineration-based technologies. Water emissions, which are generally easier to monitor and control prior to their release than air emissions, would probably be greater.

Public/Stakeholder Involvement

Nonincineration-based disposal methods are likely to be more acceptable to many segments of the public, and public support could decrease the legal and regulatory delays before a nonincineration method could be in operation. Given the history of public reaction to the stockpile disposal program and the Army's public commitments on restricting the use of any stockpile facility, including nonincineration facilities, a well designed public involvement program to explore acceptability for use of stockpile facilities would be essential prior to any Army decisions. As discussed in Chapter 3, a public involvement program will also be appropriate if commercial facilities using nonincineration technology are being considered for CAIS disposal.

Programmatic Considerations

Incineration-based disposal methods are widely available in industry and at Army facilities. Far fewer nonincineration-based facilities could be used for CAIS disposal. Therefore, the selection of a commercial nonincineration facility or the development of a new nonincineration-based technology by the Army could lead to significant delays.

6

Conclusions and Recommendations

Based on the preceding analysis (summarized in Table 6-1), the committee developed a number of conclusions and recommendations. Some are general in nature, and some are related to a particular disposal option. In general, the committee found that much of the legal and regulatory burden associated with transporting and disposing of CAIS can be reduced, which would accelerate the disposal program, without significantly endangering public safety.

CLASSIFICATION AND REGULATION OF CAIS FOR TRANSPORT AND DISPOSAL

The conclusions and recommendations on classification and regulation of CAIS apply to all the disposal alternatives.

Conclusion 1. If existing Army policies and regulations, as well as U.S. laws and their interpretations, were clarified and made more internally consistent, CAIS disposal would be simplified and the number of disposal alternatives would be increased without compromising public safety. A consistent approach to regulating CAIS would be to classify the CAIS *set* or individual *items* from a set as a characteristic hazardous waste rather than as chemical warfare materiel or chemical agent. This approach is consistent with historical practice in environmental regulation, in which a waste is classified on the basis of the amount of chemical constituents it contains and the potential risks it poses. If CAIS sets and items were classified as a characteristic hazardous waste, this would not (and should not) set a precedent for reclassifying any of their chemical constituents, such as sulfur mustard, that are classified as chemical warfare agents or chemical warfare materiel when in other configurations.

Conclusion 1a. CAIS can be safely transported and handled if the best industrial practices for highly hazardous materials are used for packaging, handling, worker safety, monitoring, plant inspections, and audits, particularly if these practices are used in conjunction with the Army's experience in handling CAIS materials. Because either specialized commercial or Army-specific facilities and equipment could be used for transport and disposal, much of the present regulatory burden and Army bureaucracy surrounding the handling, transport, and disposal of CAIS items seems to be unnecessary.

Conclusion 1b. For the purposes of transportation and disposal, CAIS containing mustard and lewisite could safely be classified as hazardous waste and not as chemical

TABLE 6-1 Summary Evaluation for all CAIS Disposal Options[a]

	Commercial Incineration	Baseline, Mobile RRS	Fixed RRS	Nonincineration[b]
Technology				
Process reliability and effectiveness	Well proven for mustard; arsenic-containing agents may require special treatment.	Neutralization process is proven; reliability and effectiveness appear to be high; some issues remain unresolved.	Neutralization process is proven; reliability and effectiveness appear to be high; some issues remain unresolved.	Neutralization proven during stockpile program development; other processes are under development or unproven.
Technical maturity	Mature, but process modifications may be required.	Process chemistry is mature; RRS system is being tested.	Process chemistry is mature; RRS system is being tested.	Some commercial processes exist; agent-specific treatment processes are under development.
Monitoring and disposal of process effluents	Committee recommends continuous air monitoring in receive/unpack areas; public may require "hold and test" monitoring of emissions and effluents.	Liquid process wastes must be packaged, transported, and treated; liquid wastes must be characterized to ensure safe disposal.	Liquid process wastes must be packaged, transported, and treated; liquid wastes must be characterized to ensure safe disposal.	Unknown, but no monitoring is expected beyond routine analysis of residual wastes prior to release; agent monitors could be an added cost at a commercial facility.
Laws and Regulations				
Consistency with present laws, regulations, and treaties	Non-Army disposal requires legal/regulatory relief, clarification, or flexibility; some facility permit modifications may be required by EPA.	State-by-state and site-specific RCRA permitting could lead to significant delays and costs.	RRS permitting requirements by states and EPA are reduced; approvals for CAIS transportation by U.S. Dept. of Health and Human Services are increased.	Requires legal/regulatory relief, clarification, or flexibility; some facility permit modifications may be required.
Costs				
Permitting	Permit modifications, if required, may add cost; permit restrictions may affect processing costs.	Permit required for each state in which RRS is used; RCRA permit required to store CAIS for more than 90 days	Several operating permits necessary for RRSs; permits may limit use to in-state or known CAIS items; permits and transportation plans are required to ship CAIS.	Same permits required for all CAIS disposal alternatives.
Indemnification	A potential added cost to the Army or the facility.	None	None	A potential added cost to the Army or facility.
Facility modifications	Monitoring and other modifications may increase costs.	None	None	Monitoring and other modifications may add cost.
Transportation	Transportation to commercial sites may increase cost; escorts may be required; not clear who pays for handling, characterization, and transport.	Transportability of RRS is a major advantage, but transporting and staffing costs are considerable; treatment of liquid wastes at commercial facilities adds to cost.	Transporting CAIS to RRS with escorts is an added cost, but field staffing costs are lower; treatment of liquid wastes at commercial facilities adds cost.	Transportation to commercial sites may add cost; military escorts may be required; not clear how handling, characterization, and transportation costs are funded.

Table 6-1 continued

	Commercial Incineration	Baseline, Mobile RRS	Fixed RRS	Nonincineration[b]
Processing operations	Dedicated processing of CAIS could be costly; CAIS packaging could be an added cost.	Estimated costs of processing (site preparation, set-up, operations, closure) are high; large staff and overhead required; support costs of RRS between deployments required.	No site preparation or closure costs; in-field costs of characterizing, separating, and packaging CAIS would be incurred.	CAIS packaging would add costs; dedicated processing of CAIS could be costly.
Indirect Costs	Hidden indirect costs (overhead, administration, maintenance).	Cost recovery for design and construction; usage fees.	Cost recovery for RRS design and construction; usage fees.	Hidden indirect costs (overhead, administration, maintenance).
Environmental Impacts, Worker/Public Safety, and Risks				
Environmental impact	Air emissions minimized by facility design.	Will be assessed during RRS test program and initial permitting.	Will be assessed during RRS test program and initial permitting.	Air/water emissions minimized by nature of process.
Worker safety	Monitoring in receive/unpack/areas needed; training and protective equipment for hazardous waste handling is needed if not already adequate; hazards seem manageable for facilities permitted for hazardous wastes of comparable toxicity.	Will be assessed during RRS test program and initial permitting.	Will be assessed during RRS test program and initial permitting.	Will be assessed after technology identified and tested.
Public safety	Impacts on public safety controlled by government regulations.	Will be assessed during RRS test program and initial permitting.	Will be assessed during RRS test program and initial permitting; transportation to fixed RRS must also be assessed.	Will be assessed after technology identified and tested.
Risk analysis[c]	Risks generally known and understood for commercial facilities; CAIS chemicals seem similar to other hazardous chemicals currently being incinerated; risks to workers in receive/unpack areas should be analyzed.	Essentially covered in design and development of procedures, costs, etc.; risks of disposition of neutralized wastes unknown but less of a concern.	Essentially covered in design and development of procedures, costs, etc.; risks of disposition of neutralized wastes unknown but less of a concern.	Potential risks should be considered in the specification of any treatment option, especially for storage and handling of CAIS items.

Table 6-1 continued

	Commercial Incineration	Baseline, Mobile RRS	Fixed RRS	Nonincineration[b]
Public/Stakeholder Involvement	Perceived public health issue concerning chronic risks from incinerator emissions; transporting large numbers of CAIS, or CAIS types containing large volumes of agent, may be an issue; priority should be on allocating resources for public involvement.	A mobile facility is likely to be more acceptable than a permanent, fixed facility; however, incineration of RRS wastes is strongly opposed by some segments of the public.	Incineration of RRS wastes is strongly opposed by some segments of the public; Army should seek public approval of RRS sites.	Nonincineration-based methods likely to be more acceptable to many members of the public.
Programmatic Considerations				
Schedule	Could allow prompt disposal of small recoveries of CAIS, but public resistance and regulatory treatment may lead to significant delays.	Movement and permitting of RRS could cause delays.	CAIS transportation approvals may cause limited delays.	Significant delays possible during technology development or identification; more rapid disposal schedule possible once available.
Funding	Liability and contractual issues could increase costs.	Operational funding requirements are significant.	Operational funding required.	Significant funds required for any technology development program.
Organizations	Corporate commitment is a significant unknown.	Movement of RRS would require coordination.	RRS sites would have to be approved by base commanders.	Corporate commitment is unknown.

[a] Any CAIS disposal option would have to address the CAIS container, the agents contained therein, and any resultant waste products.

[b] For example, lewisite could be treated by the technology already in use in the Army's CAMDS (Utah) facility for destruction of bulk lewisite (the Canadian Swiftsure process—neutralization followed by immobilization of the arsenic-containing products in a cement-like matrix that is subsequently disposed of in a landfill). Sulfur mustard could be treated by the technology to be used in the chemical stockpile disposal facility being built at Aberdeen Proving Ground (Maryland) for destruction of sulfur mustard in ton containers (neutralization, followed by biodegradation), or technologies at other commercial facilities could potentially be used.

[c] Risk analysis includes identifying of hazards, understanding the risks, identifying risk control measures, and putting risks into context. The initial discovery of CAIS items, particularly by untrained members of the public, seems to pose the greatest risks. However, the committee's analysis begins at the point of CAIS recovery.

warfare materiel. The reclassification would greatly reduce the costs of transportation and disposal and would substantially increase the feasibility of CAIS disposal. This change should have no impact on the safety of CAIS recovery, transportation, or disposal operations for the following reasons:

- CAIS contain no explosives.
- The chemicals in recovered or stored CAIS that are currently interpreted in Defense Department guidance as chemical warfare agents are sulfur mustard and lewisite. These chemicals are considered to have relatively high inherent hazard (at the high end of the range of hazards presented by hazardous industrial chemicals). Nevertheless, the risk posed by proper treatment of small quantities of these is less than the risk posed by the larger quantities of highly hazardous industrial chemicals that are already handled by the chemical industry and commercial hazardous waste treatment facilities. Although some CAIS configurations contain potentially lethal quantities of chemicals, the risks to the public and workers in handling CAIS can be controlled to protect human health.
- Most CAIS (except for two types that contain several liters of agent per set) contain relatively small quantities of chemical ingredients, often in dilute forms.

Recommendation 1. The Army should present a plan to Congress describing how it will work with regulators, other appropriate decision makers, and stakeholders to clarify the regulatory status of Chemical Agent Identification Sets (CAIS), either through separate legislation (as part of 50 U.S.C. section 1512) or by other appropriate means. A range of stakeholders and public groups should be included in this process to ensure that this proposal to clarify regulations is presented in a forthright manner. In particular, the Army should inform the public that CAIS items contain chemical warfare agents and should be explicit about the technologies that would be used for commercial disposal. This plan should be part of the Army's overall program for CAIS disposal and should address ancillary issues, such as the implications of the Chemical Weapons Convention. One alternative that should be explored through this process is the feasibility of classifying complete CAIS sets or items from sets as a characteristic hazardous waste.

COMMERCIAL INCINERATION

Conclusion 2. Even though commercial incineration seems technically feasible and may offer cost and time savings compared to the RRS, many hurdles would have to be overcome. Not the least is ensuring that commercial incineration of CAIS is acceptable to the public.

Recommendation 2. If the Army and its stakeholders cannot agree that the commercial incineration of CAIS is practical, the Army should expand its inquiry to include other disposal alternatives, such as nonincineration disposal methods, in either Army or commercial facilities, using technologies that have already been used in operational, permitted facilities or are scheduled to be demonstrated.

Conclusion 3. It is technically feasible to dispose of all CAIS items in commercial hazardous waste incineration facilities that have a permit specifically addressing wastes containing arsenic and that operate at the highest level of destruction and removal efficiencies for organic compounds. An example would be a permit specifying destruction and removal efficiencies similar to those required for commercial incineration facilities permitted to treat nitrogen mustard, polychlorinated biphenyls, or dioxins. Disposal in these commercial incineration facilities can be safe, reliable, and effective. The committee anticipates that a thorough and well-documented comparison of risk components will show that the risk to the public from the incineration of *smaller quantities* of CAIS items is lower than the risk from the routine incineration of *larger quantities* of highly toxic industrial chemicals. With appropriate process controls and monitoring, as discussed in this report, the committee also anticipates that risks to workers from incineration of CAIS items will be no greater than the risks from other commercially incinerated materials that are routinely handled in these facilities.

Recommendation 3. To provide a documented evaluation of the environmental and worker/public safety issues involved in the commercial incineration of CAIS, the Army should prepare a report that compares the relative risks to workers and the public of incinerating CAIS items with the risks to workers and the public of incinerating highly hazardous industrial chemicals at any facility proposed for CAIS disposal. Among the components of risk that should be documented are (1) the toxicity of chemical agents in CAIS (mustard and lewisite) relative to highly hazardous industrial chemicals (e.g., agent-contaminated materials, highly toxic industrial chemicals, polychlorinated biphenyls, medical wastes, and other hazardous military wastes) that are routinely destroyed in commercial incineration facilities; (2) the anticipated annual volumes of agents in CAIS to be disposed of, compared with the annual volumes of highly hazardous industrial chemicals that are currently being commercially incinerated; and (3) the Environmental Protection Agency's "incinerability" classifications of chemicals in CAIS and highly hazardous industrial chemicals.

Conclusion 4. By law, chemical warfare agent disposal facilities are required to provide maximum protection of the public, workers, and the environment. However, the term "maximally safe" is not clearly defined in the statute or in Army regulations and guidance documents.

Recommendation 4. Either the Army, the U.S. Department of Defense, or Congress should clarify the interpretation of "maximally safe" to ensure that it can be applied consistently in different situations. For the transportation and handling risks, the role of feasibility in determining what is maximally safe should be incorporated through the use of regulatory concepts such as ALARA (as low as reasonably achievable) or ALARP (as low as reasonably practicable). For the risks from emissions and discharges, the well-established regulatory policy for managing waste disposal risks should be applied. For all risks, a risk management approach should be used to ensure that appropriate controls are identified and evaluated.

Conclusion 5. The Army and its contractor conducted a preliminary analysis of the technical feasibility of commercial disposal of CAIS items at selected sites by incineration. The analysis was based on destruction of similar materials, and no trial burns were conducted. Sulfur mustard, the major chemical of concern in CAIS items, has been successfully destroyed via incineration and chemical neutralization. Lewisite, an arsenic-based material, has also been destroyed successfully, but, if it is incinerated, special scrubbing equipment may be required to meet regulatory limits on arsenic emissions. Although the committee does not know whether the facilities surveyed by the

Army could handle arsenic-based materials, there are commercial incinerators that have permits allowing them to treat wastes containing arsenic. Characterization of incoming wastes (for compliance with a facility permit), monitoring of destruction removal efficiencies and emissions (particularly arsenic), and special handling (unless CAIS overpacks containing mustard or lewisite could be fed directly into the disposal equipment) may be required at commercial facilities. These requirements, combined with possible process and permit modifications, could be major economic and technical hurdles for commercial facilities.

Recommendation 5. The Army should develop a stronger technical basis for its conclusion that commercial incineration of items from Chemical Agent Identification Sets (CAIS) is technically feasible (e.g., by determining if anything unique about CAIS disposal would preclude commercial incineration). The Army should also provide side-by-side data showing the destruction kinetics of CAIS and highly hazardous chemicals already being destroyed in commercial facilities. The data should be consistent with the conditions at state-of-the-art commercial facilities (i.e., facilities permitted to handle hazardous chemicals, such as polychlorinated biphenyls, dioxins, or nitrogen mustard).

Conclusion 6. A preliminary cost estimate developed by the Army and its contractor showed that commercial incineration of CAIS items could yield substantial cost savings compared with the RRS option. However, a number of items either were not included or were not adequately discussed in this preliminary cost estimate (e.g., permit modifications, transportation of CAIS items, packaging, agent monitoring and other facility modifications, and staff training). In contrast to this optimistic estimate, the projected costs of the Army's baseline approach (i.e., the mobile RRS) seem overly conservative. Furthermore, the preliminary estimate did not include programmatic issues for the commercial incineration option. If the commercial option is pursued, issues of corporate commitment, legal liability, public notification requirements, and contractual matters could arise.

The Army's cost estimate for commercial incineration was two orders of magnitude lower than the estimate for the RRS, which implies a potential for significant savings even after accounting for the costs not included in the estimate. However, given the potential regulatory problems, public concerns, and liability barriers, the Army may have to remove barriers before commercial firms will undertake CAIS disposal.

Recommendation 6. The committee concurs with the Army's finding that a comparative cost analysis of commercial facilities with the options for the Rapid Response System should be conducted. The existing analysis provided by the Army is inadequate for this purpose. The cost analysis should be more detailed and, to the extent possible, should include all relevant costs so that accurate comparisons can be made.

Conclusion 7. The Army's report to Congress did not include a risk assessment for the commercial (incineration) disposal option; in fact, it did not discuss the risks at all. However, because phosgene and other CAIS ingredients are routinely used and disposed of in the chemical industry in much larger quantities than occur in CAIS, it seems reasonable to assume that the risks during CAIS disposal could be controlled. The Army's risk evaluation framework for the Chemical Stockpile Disposal Program could be adapted for application to CAIS disposal options.

Recommendation 7. To characterize the risks of the commercial incineration option, the Army should conduct a risk evaluation using various hazard identification and evaluation methodologies, as appropriate. The evaluation of risks should include risks from delays, from transporting CAIS to a commercial facility, from handling CAIS in a commercial facility, and from treating CAIS disposal effluents. Worker safety during CAIS disposal should be evaluated using objective safety criteria to determine the degree of specialized personal protective gear, workplace monitoring equipment, and/or specialized training that may be necessary. If the evaluation indicates risks to workers or the public that appear to warrant further risk control measures, then more detailed risk assessments may be helpful. Commercial operations for CAIS disposal should use procedures and provide protection equivalent to the safety practices that have been determined to be necessary in military installations that handle CAIS.

Conclusion 8. The commercial incineration option may encounter public opposition by various groups, which could lead to schedule delays and added costs similar to those experienced by the Chemical Stockpile Disposal Program. Unfortunately, the Army's report to Congress did not include a detailed analysis of public acceptability issues—including how CAIS disposal would be related to the overall strategy for the disposal of non-stockpile materiel from the public's perspective. Instead, the report focused on cost, technical efficiency, and legal issues. Past experience has shown that focusing on these issues alone does not ensure public acceptability. Whichever option the Army favors, considerable staffing and funding for public involvement activities will be required to facilitate selection of an option that is both technically sound and acceptable to the public.

Recommendation 8. If the commercial disposal option is pursued, the Army should carefully assess the public acceptability challenges of commercial incineration and ensure that the necessary resources and staff (skills, experience, and number) are available to develop and implement an effective public involvement program. This program should be coordinated with similar activities Army-wide, particularly activities of the Chemical Stockpile Disposal Program, to ensure that the approaches to public involvement are consistent.

RAPID RESPONSE SYSTEM

The Army's current plans for CAIS disposal are based on the use of a transportable RRS, which is currently being tested. The committee found that the RRS, in both the baseline, mobile configuration and the fixed mode, offers advantages in mobility and simplicity of operation (important attributes from the public's perspective), as well as the capability to characterize, separate, and repackage individual CAIS items. However, the committee also found that operational costs, permitting requirements, and follow-on treatment of RRS wastes are issues that must be addressed prior to using either RRS configuration.

Mobile Rapid Response System

Conclusion 9. Although some national and regional stakeholder groups have endorsed the concept of a mobile facility, a number of unresolved issues will make the disposal of

CAIS via the mobile RRS difficult. Preliminary cost estimates indicate that RRS deployments will be expensive and more time consuming than the Army originally envisioned. For example, state-by-state permit requirements will hinder the rapid use of the RRS, and processing and transport costs in the Army's estimate seem unusually high. The RRS neutralization scheme seems viable as a preliminary processing step, although the entire RRS has not yet been fully tested as a system, and issues surrounding the monitoring and subsequent disposal of process effluents, in particular the use of incineration for treating RRS wastes, have not been completely resolved.

Recommendation 9. As the Army begins initial testing of the Rapid Response System (RRS), it should critically examine a number of unresolved issues, including site-specific permitting requirements, monitoring, public involvement, and the disposal of process effluents. These issues should be resolved prior to the operational deployment of the RRS.

Conclusion 10. Only two sites have permits that would allow long-term storage of CAIS prior to the arrival of an RRS: Deseret Chemical Depot (Utah) and Pine Bluff Arsenal (Arkansas). Both sites have occasionally placed restrictions on the receipt of CAIS items. Regulatory approval for transporting CAIS items across state lines to these sites will also affect disposal costs and schedules. The procedural and regulatory requisites for transportation of CAIS could be simplified by preparing a generic plan or template with wording appropriate for all situations, such as descriptions of relevant regulations, the mode of transport to be used, handling procedures, and so on. This template could include blanks for situation-specific details, such as the locations from and to which CAIS are to be transported, the specific CAIS materials to be moved, and situation-specific risks to be addressed.

Recommendation 10. The Project Manager for Non-Stockpile Chemical Materiel should work with the Deseret Chemical Depot (Utah) and Pine Bluff Arsenal (Arkansas) storage facilities to clarify their acceptance criteria for Chemical Agent Identification Sets or items from them. The project manager should also consider developing alternative storage facilities in case these facilities become temporarily unavailable. The Army should work with regulators to reduce the time and administrative costs of developing transportation plans, recognizing that portions of these plans will necessarily be site-specific.

Fixed Rapid Response System

Conclusion 11. Disposal of CAIS by means of the fixed RRS approach seems to offer potential cost savings by reducing the requirements for site-specific disposal permits and facility transportation. However, transporting CAIS to a fixed RRS will require regulatory approval and may be less attractive to some members of the public than a mobile facility. Regulatory costs could be significant unless the Army can obtain generic transportation permits or other forms of administrative relief.

Recommendation 11. If the fixed (regional) option for the Rapid Response System is pursued, the Army must move quickly to engage base commanders, regulators, and public and stakeholder groups in exploring the details of this approach, including the disposal of process effluents and the locations of the fixed facilities.

NONINCINERATION-BASED OPTIONS

Conclusion 12. The Army's Alternative Technologies and Approaches Program and Assembled Chemical Weapons Assessment Program have identified several nonincineration technologies for the disposal of chemical warfare agents, including sulfur mustard and possibly lewisite. These processes may be more acceptable to the public than either commercial incineration or neutralization of CAIS materiel in the RRS followed by incineration. Nonincineration processes might be implemented in either commercial or Army-owned facilities. However, the absence of economic incentives for commercial firms to make process and regulatory modifications may preclude the use of commercial facilities.

Recommendation 12. The Army should evaluate the technical feasibility of using nonincineration processes for destroying Chemical Agent Identification Sets and process effluents. The Army should also consider methods of identifying and overcoming institutional, regulatory, and economic barriers to the development of commercial nonincineration facilities.

Conclusion 13. The disposal of CAIS in Army facilities that use nonincineration methods of destruction could offer a low-cost, maximally safe option, if CAIS disposal can be conducted as part of the normal, planned operations of these facilities. The technology being used by the Chemical Agent Munitions Disposal System may be appropriate for the disposal of CAIS items containing lewisite. The neutralization-based technology planned for the facility at Aberdeen Proving Ground may be appropriate for the disposal of CAIS items containing mustard. The Army has explicitly promised concerned stakeholders not to seek to alter the federal law prohibiting the use of chemical stockpile disposal facilities for the disposal of other wastes, including CAIS. Therefore, public resistance and current legal restrictions on additional uses of the stockpile facilities may make their use for CAIS disposal impossible. Nevertheless, the use of nonincineration-based disposal technologies like those at existing or planned Army facilities appears to be a technically and economically attractive option for the disposal of CAIS containing mustard or lewisite, provided affected communities agree and are involved in establishing the conditions for use of the facilities.

Recommendation 13. Congress should consider revising the legal restrictions on the use of stockpile disposal facilities to allow the disposal of Chemical Agent Identification Sets (CAIS) at appropriate nonincineration-based facilities, at least where the local community agrees to short-term use of a facility to dispose of limited amounts of recovered and stored CAIS materials. At the same time, the Army should explore the use of nonincineration-based technologies for CAIS disposal and should engage the affected public and stakeholders at sites that will use these technologies in exploring the acceptability of this alternative.

A PATH FORWARD

The committee found that, if legal and regulatory burdens can be reduced, the CAIS disposal program could be accelerated safely, reliably, and effectively. However, implementation would require changes in current law and policy, with the advice and consent of the public.

Although the committee believes that incineration of CAIS under controlled conditions is technically acceptable, some members of the public have expressed strong opposition to incineration. Based on experience with other disposal programs and the committee's interactions with concerned public groups, the committee expects that the public may be more accepting of disposal technologies that are not based on incineration.

Summary Conclusion. All of the alternatives for disposing of CAIS evaluated by the committee have advantages and disadvantages. Although the approach, or approaches, will ultimately be selected by the Army, the committee believes the Army can take several steps to expand its options. As the Army moves forward, it will be vital that a range of public and other stakeholder groups be actively involved in decision making. The committee believes that consideration of the perspectives of these groups on risk, economic implications, and other aspects of CAIS disposal options will contribute significantly to better decisions.

Summary Recommendation. The Army should take the following actions to expand its options for cost-effective disposal of Chemical Agent Identification Sets (CAIS) without decreasing safety or increasing the risks to workers, the public, or the environment:

- The Army should reconsider its interpretation of CAIS as chemical warfare materiel under U.S.C. section 1512. If the Army decides it cannot change its interpretation, then Congress should consider amending the legislation so that CAIS sets or items from CAIS can be regulated as hazardous waste under the Resource Conservation and Recovery Act.
- The Army should promote the development of nonincineration technologies for CAIS disposal.
- The Army should develop, review with stakeholders, and then implement a written plan for public involvement designed to reach a range of stakeholders and concerned groups, including affected communities and tribal nations, state and federal regulators, concerned national and regional groups, and representatives of the waste disposal industry.
- In states with a chemical stockpile disposal facility, the Army should engage the affected communities in a discussion of alternatives, including the potential use of the stockpile facility for CAIS disposal. If a community agrees to consider using the stockpile facility (and only if it agrees), the Army should pursue that option with the full involvement of the community, including establishing specific conditions for the use of the facility. If the community agrees, which may be more feasible at facilities that use nonincineration technologies, the law prohibiting the use of chemical stockpile disposal facilities for any other purpose would have to be modified to allow CAIS disposal.

An important current capability of the RRS is that it can characterize, separate, and repackage individual CAIS items. However, because of the inherent permitting problems and high costs of the mobile RRS option, the Army should aggressively pursue other options while continuing to implement the RRS.

References

Amr, A., A. Goldfarb, S. Haus, L. Hourcle, M. Simmons, A. Talib, D. Tripler, R. Wassmann, and A. Wusterbath. 1998. Preliminary Assessment of the Commercial Viability for CAIS Treatment and Disposal. MTR-1998-5. McLean, Va.: Mitretek Systems.

Bradbury, J.A., K.M. Branch, J.H. Heerwagen, and E.B. Liebow. 1994. Community Viewpoints of the Chemical Stockpile Disposal Program. Washington, D.C.: Battelle Pacific Northwest Laboratories.

Brankowitz, W. 1998. Presentation by William Brankowitz, Office of the Project Manager, Non-Stockpile Chemical Materiel, to the Committee on Review and Evaluation of the Army Non-Stockpile Chemical Materiel Disposal Program, August 18, 1998, National Research Council, Washington, D.C.

Brankowitz, W., M. Witt, J. Ursillo, and J. Pantleo. 1983. Disposal of Chemical Agent Identification Sets at Rocky Mountain Arsenal, Colorado. Final Report, Vol. 1–3. Report No. DRTH-IS-FR-83203. August 1983. Aberdeen Proving Ground, Md.: U.S. Army Toxic and Hazardous Materials Agency.

Brooks, M.E., and G.A. Parker. 1979. Incineration/Pyrolysis of Several Agents and Related Chemical Materials Contained in Identification Sets. ARCSL-TR-79040. Aberdeen Proving Ground, Md.: Program Manager for Chemical Demilitarization.

CDC (Centers for Disease Control). 1988. Final recommendations for protecting the health and safety against potential adverse effects of long-term exposure to low doses of agents GA, GB, VX, mustard agent (H, HD, T), and lewisite (L). Centers for Disease Control, U.S. Department of Health and Human Services. Federal Register 53(50): 8504–8507.

Curlee, T.R., S.M. Schexnayder, D.P. Vogt, A.K. Wolfe, M.P. Kelsay, and D.L. Feldman. 1994. Waste to Energy in the United States: A Social and Economic Assessment. Westport, Conn.: Quorum Books.

Defense Environmental Alert. 1999. Citizens urge Army to dismiss incineration for non-stockpile weapons. Defense Environment Alert, March 9, 1999, pp. 6–7.

Dempsey, C.R., and E.T. Oppelt. 1993. Incineration of hazardous wastes: a critical review update. Air and Water 43: 25–73.

DiMichele, B. 1999. Response tackles hazards of chemical ordnance. Huntsville, Ala.: U.S. Army Engineering and Support Center, Huntsville Center Public Affairs Office. Available on line at URL: <www.hnd.ace.army.mil/oew/news/9raritan.html>.

DOT (U.S. Department of Transportation). 1996. North American Emergency Response Guidebook. Research and Special Projects Administration, Department of Transportation. Washington, D.C.: U.S. Department of Transportation.

Edelstein, M.R. 1988. Contaminated Communities: The Social and Psychological Impacts of Residential Toxic Exposure. Boulder, Colo.: Westview Press.

EPA. 1989. Handbook: Guidance on Setting Permit Conditions and Reporting Trial Burn Results. Vol. 2 of the Hazardous Waste Incineration Guidance Series. Office of

Research and Development. EPA/625/6-89/019. Washington D.C.: Environmental Protection Agency.

EPA. 1998a. Communication from Region IX staff member. Defense Environment Alert, January 26, 1999, p. 8.

EPA. 1998b. Human Health Risk Assessment Protocol for Hazardous Waste Combustion Facilities. Vol. 1. Office of Solid Waste and Emergency Response. Washington, D.C.: Environmental Protection Agency.

EPA. 1999. Personal communication from Norma J. Abdul-Malik, Office of Solid Waste, Permits and State Programs Division, to staff of the Committee on Review and Evaluation of the Army Non-Stockpile Chemical Materiel Disposal Program, February 16, 1999.

Fatz, R.J. 1997. Memorandum from Raymond J. Fatz, Deputy Assistant Secretary of the Army, to distribution. Subject: Interim Guidance for Biological Warfare Materiel (BWM) and Non-Stockpile Chemical Warfare Materiel (CWM) Response Activities. September 5, 1997.

Fiorino, D.J. 1990. Citizen participation and environmental risk: a survey of institutional mechanisms. Science, Technology, and Human Values 15: 226–243.

Freudenberg, N. 1994. Citizen action for environmental health: report on a survey of community organizations. American Journal of Public Health 74: 444–448.

Ginsburg, R. 1992. Beyond the Rush to Burn: Alternatives to Hazardous Waste Incineration. Boston: National Toxics Campaign Fund.

Hunter, S., and K.M. Leyden. 1995. Beyond NIMBY: Explaining opposition to hazardous waste facilities. Policy Studies Journal 23(4): 601–619.

IOM (Institute of Medicine). 1993. Veterans at Risk: The Health Effects of Mustard Gas and Lewisite. Committee on the Survey of the Health Effects of Mustard Gas and Lewisite, Institute of Medicine. Washington, D.C.: National Academy Press.

Kasperson, R.E., D. Golding, and S. Tuler, 1992. Social distrust as a factor in siting hazardous facilities and communicating risk. Journal of Social Issues 48(4): 161–187.

Kentucky Environmental Foundation. 1998. Public Health and Chemical Weapons Incineration. Berea, Ky.: Kentucky Environmental Foundation.

Kovalyev, N.N. 1997. Personal communication to G. W. Parshall, member of the Committee on Review and Evaluation of the Army Non-Stockpile Chemical Materiel Disposal Program, at a meeting of the Monterey–Moscow Study Group on Russian Chemical Disarmament, Monterey, California, February 25, 1997.

Kummling, K.E., E.A. Chisholm, and F.T. Arnold. 1999. Application of Gas-Phase Chemical Reduction for Chemical Weapons Demilitarisation. Presented to the DERA Chemical Weapons Demilitarisation Conference, Vienna, Austria, June 8, 1999.

Lee, J.W., R.W. Ross, R.H. Vocque, J.W. Lewis, and L.R. Waterland. 1987. Distribution of volatile trace elements in emissions and residuals from pilot-scale liquid injection incineration. Pp. 254–261, 524, in Proceedings of the 13th Annual Research Symposium. EPA Report No. 600/9-87/015. Washington, D.C.: Environmental Protection Agency.

Lee, J.W., W.E. Whitworth, and I.R. Waterland. 1992. Pilot-Scale Evaluation of the Thermal Stability POHC Incinerability Ranking. EPA/600/SR-92/065. May 1992. Risk Reduction Engineering Laboratory. Washington D.C.: Environmental Protection Agency.

Libby, E. 1999. Personal communication from Edmund Libby, Project Manager for Non-Stockpile Chemical Materiel, to staff of the Committee on Review and Evaluation of the Army Non-Stockpile Chemical Materiel Disposal Program, August 19, 1998.

Lucas, S.V. 1997. Development and Performance Testing of a Chemical Analysis Method for Sulfur Mustard (HD), Nitrogen Mustard (HN-1) and Lewisite (L) in

Rapid Response System (RRS) Neutralization Solutions. Columbus, Ohio: Battelle Memorial Institute.

Maggio, C. 1998. Lewisite Demilitarization. Presentation by Cheryl Maggio, U.S. Army, Office of the Product Manager, Alternative Technologies and Approaches, Office of the Program Manager for Chemical Demilitarization, to the Committee on Review and Evaluation of the Army Chemical Stockpile Disposal Program, Woods Hole, Massachusetts, June 25, 1998.

Martens, H. 1998. Recovered old arsenical and mustard munitions in Germany. Pp. 33–78 in Arsenic and Old Mustard: Chemical Problems in the Destruction of Old Arsenical and "Mustard" Munitions, J. F. Bunnett and M. Mikolaczyk, eds. Proceedings of a NATO Workshop, March 1996, Lodz, Poland. Dordrecht, Netherlands: Kluwer Academic Publishers.

McAndless, J.M., V. Fedor, and T. Kinderwater. 1992. Destruction and Waste Treatment Methods Used in a Chemical Agent Disposal Project. Ralston, Alberta, Canada: Defense Research Establishment Suffield. (Available from National Technical Information Service as Report No. AD A259 689.)

Mitre. 1993. Summary Evaluation of the Johnston Atoll Chemical Agent Disposal System Operational Verification Testing. MTR93W0000036. May 1993. McLean, Va.: Mitre Corporation.

NATO (North Atlantic Treaty Organization). 1995. Handbook on the Medical Aspects of NBC Defensive Operations. Part III. Report No. FM 8-9. Brussels: North Atlantic Treaty Organization.

NRC. 1993. Alternative Technologies for the Destruction of Chemical Agents and Munitions. Committee on Alternative Chemical Demilitarization Technologies, Board on Army Science and Technology. Washington, D.C.: National Academy Press.

NRC. 1994a. Recommendations for the Disposal of Chemical Agents and Munitions. Committee on Review and Evaluation of the Army Chemical Stockpile Disposal Program, Board on Army Science and Technology. Washington, D.C.: National Academy Press.

NRC. 1994b. Evaluation of the Johnston Atoll Chemical Agent Disposal System Operational Verification Testing. Part 2. Committee on Review and Evaluation of the Army Chemical Stockpile Disposal Program, Board on Army Science and Technology. Washington, D.C.: National Academy Press.

NRC. 1996a. Review and Evaluation of Alternative Chemical Disposal Technologies. Panel on Review and Evaluation of Alternative Chemical Disposal Technologies, Board on Army Science and Technology. Washington, D.C.: National Academy Press.

NRC. 1996b. Understanding Risk: Informing Decisions in a Democratic Society. Committee on Risk Characterization, National Research Council. Washington, D.C.: National Academy Press.

NRC. 1997. Review of Acute Human-Toxicity Estimates for Selected Chemical Warfare Agents. Committee on Toxicology, Board on Environmental Studies and Toxicology. Washington, D.C.: National Academy Press.

NRC. 1998. Using Supercritical Water Oxidation to Treat Hydrolysate from VX Neutralization. Committee on Review and Evaluation of the Army Chemical Stockpile Disposal Program, Board on Army Science and Technology. Washington, D.C.: National Academy Press.

NRC. 1999a. Carbon Filtration for Reducing Emissions from Chemical Agent Incineration. Committee on Review and Evaluation of the Army Chemical Stockpile

Disposal Program, Board on Army Science and Technology. Washington, D.C.: National Academy Press.

NRC. 1999b. Review and Evaluation of Alternative Technologies for Demilitarization of Assembled Chemical Weapons. Committee on Review and Evaluation of Alternative Technologies for Demilitarization of Assembled Chemical Weapons, Board on Army Science and Technology. Washington, D.C.: National Academy Press.

Petrov, S.V., V.I. Kholstov, V.P. Zoubriline, and N.W. Zaviolova. 1998. Practical actions of Russia on preparation for destruction of stockpiled lewisite and mustard. Pp. 79–90 in Arsenic and Old Mustard: Chemical Problems in the Destruction of Old Arsenical and "Mustard" Munitions, J. F. Bunnett and M. Mikolaczyk, eds. Proceedings of a NATO Workshop, March 1996, Lodz, Poland. Dordrecht, Netherlands: Kluwer Academic Publishers.

Proctor, N.H., and J.P. Hughes. 1978. Chemical Hazards of the Workplace. Philadelphia: J.B. Lippincott Company.

Rabe, B.G., 1994. Beyond NIMBY: Hazardous waste siting in Canada and the United States. Washington, D.C.: The Brookings Institution.

Schmauder, C. 1997. Memorandum from Craig R. Schmauder, Office of Counsel, U.S. Army Engineering and Support Center, Huntsville, Alabama, to Commander, U.S. Army Engineering and Support Center, Huntsville. Subject: Transportation of Chemical Agent Identification Sets (CAIS) to Commercial Facilities for Disposal. September 15, 1997.

Shaw, R.W., and M.J. Cullinane. 1998. Destruction of military toxic materials. Pp. 2821–2836 in Encyclopedia of Environmental Analysis and Remediation. R.A. Myers, ed. New York: John Wiley & Sons, Inc. Available on line at URL: <www.aro.army.mil/chemb/people/milremed.html>.

Slovic, P., B. Fischhoff, and S. Lichtenstein, 1979. Rating the risks. Environment 21(3): 14–39.

Spritzer, M.H., D.A. Hazlebeck, and K.W. Downey. 1995. Supercritical Water Oxidation of Chemical Agents, Solid Propellants and Other Hazardous Wastes. San Diego, Calif.: General Atomics, Inc.

Smithson, A.E. 1994. The U.S. Chemical Weapons Destruction Program: Views, Analysis, and Recommendations. Report Number 13. September 1994. Washington, D.C.: The Henry L. Stimson Center.

Szasz, A. 1994. Ecopopulism: Toxic Waste and the Movement for Environmental Justice. Minneapolis: University of Minnesota Press.

U.S. Army. 1988. Chemical Stockpile Disposal Program Final Programmatic Environmental Impact Statement. Aberdeen Proving Ground, Md.: U.S. Army Program Manager for Chemical Demilitarization.

U.S. Army. 1993. Interim Survey and Analysis Report. Prepared by the Project Manager for Non-Stockpile Chemical Materiel. April 1993. Aberdeen Proving Ground, Md.: U.S. Army Program Manager for Chemical Demilitarization.

U.S. Army. 1994. U.S. Army's Alternative Demilitarization Technology Report to Congress. Department of the Army. Aberdeen Proving Ground, Md.: U.S. Army Program Manager for Chemical Demilitarization.

U.S. Army. 1995a. Chemical Agent Identification Sets (CAIS) Information Package. Prepared by the Project Manager for Non-Stockpile Chemical Materiel. November 1995. Aberdeen Proving Ground, Md.: U.S. Army Program Manager for Chemical Demilitarization.

U.S. Army. 1995b. Recovered Chemical Warfare Material. Pp. 27–29 in Nuclear and Chemical Weapons and Materiel: Chemical Surety. Army Regulation (AR) 50-6. Effective March 1, 1995. Washington, D.C.: Headquarters, Department of the Army.

U.S. Army. 1996. Survey and Analysis Report. 2nd ed. Prepared by the Project Manager for Non-Stockpile Chemical Materiel. December 1996. Aberdeen Proving Ground, Md.: U.S. Army Program Manager for Chemical Demilitarization.

U.S. Army. 1997a. Engineering Evaluation/Cost Analysis for the Treatment and Disposal of Chemical Agent Identification Sets Recovered from the Poleline Road Disposal Area, Fort Richardson, Alaska. Prepared by the Project Manager for Non-Stockpile Chemical Materiel. May 1997. Aberdeen Proving Ground, Md.: U.S. Army Program Manager for Chemical Demilitarization.

U.S. Army. 1997b. Rapid Response System. Attachment 2. Permit application to the state of Utah. Project Manager for Non-Stockpile Chemical Materiel. Available on line at URL: <www.eq.state.ut.us/eqshw/cds/RRSpermit>.

U.S. Army. 1997c. Dichlorodimethylhydantoin Treatment of Chemical Agents in Chloroform (Red Process). Final, June 1997; Dichlorodimethylhydantoin Treatment of Sulfur Mustard (Blue Process), Final, June 1997; Dichlorodimethylhydantoin Treatment of Chemical Agents on Charcoal (CHARCOAL and CHARCOAL-L processes), Final, June 1997. Prepared by the Project Manager for Non-Stockpile Chemical Materiel. Aberdeen Proving Ground, Md.: U.S. Army Program Manager for Chemical Demilitarization.

U.S. Army 1997d. Army Regulation 385-61. The Army Chemical Agent Safety Program. February 28, 1997, unclassified. Washington D.C.: Headquarters, Department of the Army.

U.S. Army. 1998a. Report to Congress on Alternative Approaches for the Treatment and Disposal of Chemical Agent Identification Sets (CAIS). Prepared by the Project Manager for Non-Stockpile Chemical Materiel. June 1998. Aberdeen Proving Ground, Md.: U.S. Army Program Manager for Chemical Demilitarization.

U.S. Army. 1998b. Chemical Agent Identification Set Treatment Comparison Study. Final (May 1998). Prepared by the Project Manager for Non-Stockpile Chemical Materiel. Aberdeen Proving Ground, Md.: U.S. Army Program Manager for Chemical Demilitarization.

U.S. Army. 1999. Status of Agent Destruction at JACADS and TOCDF. Report for Electronic Distribution. March 28, 1999. Aberdeen Proving Ground, Md.: Chemical Stockpile Disposal Program.

Velzy Associates. 1990. Assessment of Burning Toxic Metal-Bearing Wastes in a Hazardous Waste Incinerator. Report prepared for Rohm & Haas Company and SmithKline Beecham. October 1990. Spring Hill, Pa.: Rohm & Haas Company.

Wakefield, P. 1999. Personal communication from Patrick J. Wakefield, Office of the Assistant Secretary of the Army for Acquisition, Logistics and Technology, to staff of the Committee on Review and Evaluation of the Army Non-Stockpile Chemical Materiel Disposal Program, January 29, 1999.

Walsh, E.J., R. Warland, and D.C. Smith. 1997. Don't Burn It Here: Grassroots Challenges to Trash Incinerators. University Park, Pa.: Pennsylvania State University Press.

War Department. 1944. Use of Chemical Agents and Munitions in Training. Technical Manual TM3-305. Washington, D.C.: War Department.

Wynne, B. 1992. Risk and social learning: reification to engagement. pp. 275–297 in Social Theories of Risk, S. Krimsky and D. Golding, eds., Westport, Conn.: Praeger Press.

Yang, Y-C. 1995. Chemical reactions for neutralizing chemical warfare agents. Chemistry and Industry Vol. pp. 334–337.

Yang, Y-C., J.A. Baker, and J.R. Ward. 1992. Decontamination of chemical warfare agents. Chemical Reviews 92: 1729–1743.

APPENDICES

A

Biographical Sketches of Committee Members

John B. Carberry (chair) is director of environmental technology for E.I. DuPont de Nemours and Company, where he has been employed since 1965. Since 1988, he has been involved with initiatives to advance DuPont's environmental excellence through changes in products and processes and the recycling of materials with an emphasis on reducing waste and using affordable, publicly acceptable technologies for the abatement, treatment, and remediation of environmental pollution. Mr. Carberry is chairman of the Chemical Engineering Advisory Board at Cornell University, a member of the Radioactive Waste Retrieval Technology Review Group for the U.S. Department of Energy (DOE), a member of the Pollution Prevention Program Committee of the American Chemical Society, and a member of the Committee on Industrial Environmental Performance Metrics of the National Academy of Engineering. He is a fellow of the American Institute of Chemical Engineers and holds an M.S. in chemical engineering from Cornell University and an M.B.A. from the University of Delaware.

John C. Allen is vice president, transportation practice, at ICF Kaiser. He previously held several positions at Battelle Memorial Institute, including vice president in the Transportation Division and manager of the Cambridge Operations; managerial roles in the Operations Research and Analysis Program and the Institutional and Policy Analysis Program; and transportation economist. Prior to joining Battelle in 1983, he was a transportation economist and policy analyst with the U.S. Department of Transportation. Mr. Allen has managed and participated in numerous studies involving the policy, regulatory, institutional, and safety aspects of transporting hazardous and nuclear materials. He has served on various advisory panels and has been chairman of the National Research Council (NRC) Transportation Research Board's Committee on Hazardous Materials Transportation for the past four years. He holds an M.B.A. in transportation from the University of Oregon and a B.A. in economics from Western Maryland College.

Lisa M. Bendixen is principal for safety and risk analysis at Arthur D. Little, Inc. Since joining the company in 1980, she has been involved in risk management and risk assessment studies for numerous industries. Ms. Bendixen is the secretary of the NRC Transportation Research Board's Committee on Hazardous Materials Transportation and the U.S. delegate to the International Electrotechnical Commission's working group on risk analysis. She was a member of the NRC committee that evaluated the safety of fiber drums and is past chair of the Safety Engineering and Risk Analysis Division of the American Society of Mechanical Engineers (ASME). She has been engaged in various studies on the chemical demilitarization of M55 rockets, including identifying and quantifying failure modes leading to agent release based on a generic facility design for disposal; evaluating sources of risk separating agent from energetic components within the rocket; and preparing criteria for evaluating storage, transportation, and on-site disposal options on a comparative basis. Ms. Bendixen earned an M.S. in operations research at the Massachusetts Institute of Technology.

Judith A. Bradbury is a technical manager at Battelle Pacific Northwest National Laboratory currently evaluating the effectiveness of the DOE's 12 site-specific advisory boards. Previously, she led a number of assessments for the Environmental Protection Agency (EPA) on public involvement programs and concerns about incineration and community perspectives on the U.S. Army Chemical Weapons Disposal Program. Dr. Bradbury is a member of the International Association of Public Participation Practitioners. She earned a B.S. in sociology from the London School of Economics, an M.A. in public affairs from Indiana University of Pennsylvania, and a Ph.D. in public and international affairs from the University of Pittsburgh.

Martin C. Edelson is the director of the Environmental Technology Development Program at the Ames Laboratory and adjunct associate professor of nuclear engineering at Iowa State University. He has held a number of research positions at Ames Laboratory since 1977. His research interests include risk communication and the development of laser-based methods for materials processing and characterization. Dr. Edelson was a member of the Munitions Working Group, the DOE Laboratory Directors' Environmental and Occupational/Public Health Standards Steering Group, and the DOE Strategic Laboratory Council. Dr. Edelson is a technical editor of *Risk Excellence Notes*, a new publication funded by the DOE Center for Risk Excellence. He earned a B.S. in chemistry and an M.A. in physical chemistry from City College of New York and a Ph.D. in physical chemistry from the University of Oregon.

Sidney J. Green is chief executive officer of TerraTek in Salt Lake City, a geotechnical research and services firm focused on natural resource recovery, civil engineering, and defense problems. Previously, he worked at General Motors and at the Westinghouse Research Laboratory. He has an extensive background in mechanical engineering, applied mechanics, materials science, and geoscience applications and is a former member of the NRC Geotechnical Research Board. Mr. Green is a member of the National Academy of Engineering. He was named Outstanding Professional Engineer of Utah and was awarded the ASME Gold Medallion Award and the Society of Experimental Mechanics Lazan Award. Mr. Green received an M.S. in engineering mechanics from Stanford University and an M.S. from the University of Pittsburgh and a B.S. from the University of Missouri at Rolla, both in mechanical engineering.

Brigadier General Paul F. Kavanaugh (retired) is an engineering management consultant. Previously, he was the director of government programs for Rust International, Inc., and director of strategic planning for Waste Management Environmental Services. During his military service, he served with the U.S. Army Corps of Engineers, DOE, and the Defense Nuclear Agency and managed projects dealing with chemical demilitarization at Johnston Atoll. He earned a B.S. in civil engineering from Norwich University and an M.S. in civil engineering from Oklahoma State University.

Douglas M. Medville recently retired from Mitre Corporation as program leader for chemical materiel disposal and remediation. He has led many analyses of risk, process engineering, transportation, and alternative disposal technologies and has briefed the public and senior military officials on the results. Mr. Medville led the evaluation of the operational performance of the Army's chemical weapon disposal facility on Johnson

Atoll and directed an assessment of the risks, public perceptions, environmental aspects, and logistics of transporting recovered nonstockpile chemical warfare materiel to candidate storage and disposal destinations. Previously, he worked at Franklin Institute Research Laboratories and at General Electric. Mr. Medville earned a B.S. in industrial engineering and an M.S. in operations research, both from New York University.

James W. Mercer is executive vice president for HSI GeoTrans, Inc., having served as president of GeoTrans, Inc., from 1979 to 1996. Previously, he was a hydrologist at the U.S. Geological Survey. His expertise includes groundwater hydrology, multiphase flow in porous media, groundwater pollution and aquifer water quality, solute and heat transport, hazardous waste disposal, and environment remediation. Dr. Mercer has served on numerous technical committees for EPA, DOE, the U.S. Department of Defense, and NRC, including the Committee on Non-Invasive Characterization of the Shallow Subsurface for Environmental and Engineering Applications. He has also published extensively in the areas of groundwater modeling, groundwater contamination, and hazardous waste disposal. Dr. Mercer is a fellow of the American Geophysical Union. He has a Ph.D. and M.S. in geology from the University of Illinois and a B.S. in geology from Florida State University.

Winifred G. Palmer is a toxicologist with the Henry M. Jackson Foundation for the Advancement of Military Medicine and is working under a five-year grant from the U.S. Army Center for Health Promotion and Preventive Medicine. From 1989 to 1996, she was a toxicologist for the U.S. Army at both Fort Detrick and Aberdeen Proving Ground in Maryland. Her recent work has included assessments of health risks associated with chemical warfare agents, the development of a military field water quality standard for the nerve agent BZ, and studies on the bioavailability of TNT and related compounds in composts of TNT-contaminated soils. Dr. Palmer is a member of the Society of Toxicology, and her numerous publications span more than two decades of work in the field. She has a Ph.D. in biochemistry from the University of Connecticut and a B.S. in chemistry and biology from Brooklyn College.

George W. Parshall is a consultant for E.I. Du Pont de Nemours and Company, from which he retired in 1992 after a career spanning nearly 40 years, including more than 10 years as director of chemical science in the Central Research and Development Department. His expertise encompasses organic and inorganic chemistry and catalysis and conducting and supervising chemical research. Dr. Parshall is a member of the National Academy of Sciences and is a past member of the NRC Board on Chemical Science and Technology and the NRC Committee on Review and Evaluation of the Army Chemical Stockpile Disposal Program. He earned a Ph.D. in organic chemistry from the University of Illinois.

James P. Pastorick is president of Geophex UXO, Ltd., an unexploded ordnance (UXO) remediation firm based in Alexandria, Virginia, that specializes in implementing advanced geophysical UXO detection methods. He retired from the U.S. Navy as an explosive ordnance disposal officer and diver in 1989, when he began working on civilian UXO clearance projects. Prior to starting his present company, he was the senior project manager for UXO projects at UXB International, Inc., and IT Group.

William J. Walsh is an attorney and a partner in the Washington, D.C., office of Pepper Hamilton LLP. Prior to joining Pepper, he was section chief in the EPA Office of Enforcement. His legal experience encompasses environmental litigation on a broad spectrum of issues pursuant to a variety of environmental statutes, including the Resources Conservation and Recovery Act (RCRA) and the Toxic Substances Control Act. He represents trade associations, including the Biotechnology Industry Organization, in rule-making and other areas of public policy; represents individual companies in environmental actions (particularly in negotiating cost-effective remedies in pollution cases involving water, air, and hazardous waste); and advises technology developers and users on taking advantage of the incentives for, and eliminating the regulatory barriers to, the use of innovative environmental technologies. He has served on NRC committees concerned with Superfund and RCRA corrective action programs and the use of appropriate scientific groundwater models in RCRA programs and related activities. Mr. Walsh holds a J.D. from George Washington University Law School and a B.S. in physics from Manhattan College.

Ronald L. Woodfin is a staff member of the Sandia National Laboratories. He currently coordinates work on mine countermeasures and demining, including sensor development. Previously, he worked at the Naval Weapons Center, Naval Undersea Center, and at Boeing Commercial Airplane Division. He has been an invited participant at several international demining conferences. He earned a B.S. in engineering from the University of Texas and an M.S. in aeronautics and astronautics and Ph.D. in engineering mechanics from the University of Washington.

B

Committee Meetings and Other Activities[1]

FIRST MEETING

August 18–20, 1998
National Research Council
Washington, D.C.

Presentations:

Chemical Demilitarization Program
Col. Edmund Libby, Project Manager for Non-Stockpile Chemical Materiel (NSCM)

Chemical Stockpile Disposal Program
Paul Bergeron, Chemical Stockpile Disposal Program

Non-Stockpile Chemical Materiel Disposal Program
Col. Edmund Libby, NSCM

Chemical Agent Identification Sets (CAIS) Disposal Issue
William Brankowitz, NSCM

Non-Stockpile Public/Stakeholder Involvement Program
Janice Brown, NSCM

Overview of CAIS Disposal Plans and Draft Report to Congress
Robert Wassman, Mitretek

Review of Study Task and Sponsor Expectations, Scope, and Terminology
Col. Edmund Libby, NSCM

Legal and Regulatory Issues
Robert Wassman, Mitretek
Ruth Flanders, NSCM

Technical Issues
Robert Wassman, Mitretek

---

[1] The committee also gathered additional information via telephone conference calls and by other means. Details are provided on the committee's webpage <*http://www2.nas.edu/dmst/ 223a.html*>.

Costs
Arlene Wusterbarth, Mitretek

Environmental and Worker/Public Safety
Robert Wassman, Mitretek

Risks
Robert Wassman, Mitretek

Public and Stakeholder Involvement
Robert Wassman, Mitretek

Task Summary and Future Requirements
Robert Wassman, Mitretek

Programmatics
Col. Edmund Libby, NSCM

SECOND MEETING

October 28–29, 1998
National Research Council
Washington, D.C.

Presentations:

Current Program Status and Issues
Col. Edmund Libby, Project Manager for NSCM

Rapid Response System and CAIS Disposal
Larry Friedman, NSCM

Rapid Response System Costs
David Hildebrand, Science Applications International Corporation (SAIC)

CAIS Discovery and Identification Training
Steve Bird, NSCM

Army Standards for CAIS Handling and Disposal in a Commercial Disposal Facility
Steve Bird, NSCM

Quantitative/Qualitative Risks of Handling, Identification, Packaging, Storage, Transport, and Disposal of Recovered CAIS Items
Douglas Woody, SAIC

Toxicity Comparison between CAIS and Other Hazardous Wastes
Raymond Kutzman, Mitretek

APPENDIX B

The Non-Stockpile Public Outreach Program and the Status of the Programmatic Environmental Impact Statement
Col. Edmund W. Libby, NSCM
David Wilhelm, SAIC

Congressional Staff Perspectives on the Study
Jean Reed, House Committee on National Security

Environment Protection Agency (EPA) Involvement in the NSCM:

- Overall Chemical Demilitarization Program (Stockpile & Non-Stockpile)
  James Michael, EPA

- EPA "Munitions Rule"
  Kenneth Shuster, EPA

- EPA "Wastes of Concern" Project
  Dale Ruhter, EPA

Centers for Disease Control and Prevention Involvement in the NSCM
Paul Joe, Centers for Disease Control and Prevention

Department of Transportation Involvement in the NSCM
George Cushmac, U.S. Department of Transportation

Foreign Perspectives on Non-Stockpile Disposal
Robert Shaw, Army Research Office

Discussion with Members of the Non-Stockpile Forum
Ross Vincent, Sierra Club
Elizabeth Crowe, Kentucky Environmental Foundation
Ted Henry, University of Maryland

The Spring Valley Cleanup Effort
James Sweeney, D.C. Department of Health

Perspectives on Chemical Weapons Disposal
Amy Smithson, Stimson Center

Discussion with Members of the Citizens Advisory Commissions from Utah and Arkansas
Major General John Matthews (retired), Utah
Major General Don Morrow, Arkansas

THIRD MEETING

December 15–16, 1998
National Research Council
Washington, D.C.

Presentations:

Two day writing session. No Presentations.

FOURTH MEETING

June 15–17, 1999
National Research Council
Washington, D.C.

Presentations:

Evaluation of ACWA Technologies for PM-NSCM
George Bizzigotti, Mitretek

Munitions Assessment and Processing System (MAPS) June 1999 Project Overview
William Brankowitz, NSCM

Munitions Management Device (MMD) Version 1.
Alan Caplan, PMNSCM

Technology Monitoring and Evaluation of the Non-Stockpile Program
Joseph Cardito, Stone and Webster Engineering Corporation

Public Response from the Non-Stockpile Citizens Coalition
Elizabeth Crowe, Non-Stockpile Citizens Coalition

Non-Intrusive Assessment Capabilities
Ed Doyle, PMNSCM

Single CAIS Access and Neutralization Systems (SCANS)
Ed Doyle, PMNSCM

Explosive Destruction System (EDS)
Mike Duggan, PMNSCM

ACWA Program and Demonstration Update
Carl Eissner, SBCCOM

Rapid Response System Update
Larry Friedman, PMNSCM

Non-Stockpile Waste Streams/Inventory/Monitoring
John Gieseking, PMNSCM

APPENDIX B

Munitions Management Device-2
Jerry Hawks, PMNSCM

U.S. Army Non-Stockpile Chemical Materiel Project (NSCMP)
Wayne Jennings, PMNSCM

Alternative Technologies and Approaches Project Overview
J.R Ward, ATAP

SITE VISIT

August 3–4, 1999
Dugway Proving Ground, Utah
Deseret Chemical Depot, Utah

Site Team (Committee and Staff)
John Allen
Joan Berkowitz
Judith Bradbury
Martin Edelson
Sidney Green
Douglas Medville
Winifred Palmer
Ronald Woodfin

NRC Staff
Michael Clarke
Delphine Glaze
Gregory Eyring

Tour of Rapid Response System at Deseret Chemical Depot
Hosts: Michael Nuttle, Harold Oliver, Walter Levi, Brett Simms

Meeting with Utah Citizens Advisory Council (CAC)
CAC members present: Dave Ostler, Rosemary Holt, John Matthews, Dan Bauer

Tour of Supercritical Water Oxidation Facility at Dugway Proving Ground
Hosts: William Dement, Charles Donaldson, Andrew Nifsi, Beryl Schwartz, Robert Edgin, Donald Spina, Bud Salzburg, Michael Spritzer

C

Methods of Treating Non-Stockpile Chemical Materiel

This appendix describes the five types of non-stockpile chemical warfare materiel and the disposal methods being developed for their demilitarization.[1]

TYPES OF NON-STOCKPILE CHEMICAL MATERIEL

The Army has defined five categories of non-stockpile chemical warfare materiel:

- *Binary chemical weapons* form lethal chemical agents by mixing two less toxic chemicals. Army policy requires that the components of binary weapons only be loaded together into a munition immediately prior to use on the battlefield, thus forming the lethal chemical agent during flight to the target. As a result, binary components were manufactured, stored, and transported independently.
- *Buried chemical warfare materiel* includes any buried materiel. Land burial was a principal means of disposing of hazardous materials for many years, and records indicate that chemical warfare materiel was disposed of by land burial until the late 1950s.[2] (Ocean dumping was also an acceptable means of eliminating chemical warfare materiel until the late 1960s.) In most cases, the materiel was treated (burned or chemically neutralized) prior to burial. The Army is researching various methods and technologies to remediate burial sites for chemical warfare materiel.[3]

---

[1] This appendix was compiled from information on the Internet World Wide Web site for the Non-Stockpile Chemical Materiel Program. See URL: *http://www-pmcd.apgee.army.mil/text/NSCMP/index.html*.

[2] During World War I, several types of munitions were field tested and used for Army training around the country. One such munition was the WWI Livens Drum, which was typically filled with the chemical agent phosgene (choking agent) and chloropicrin (tear agent). The Livens Drum could also be used with an incendiary (flammable) or explosive fill. The Livens Drum was a short-range munition that became obsolete with the production of many long-range World War II munitions, such as the 155-mm projectile. During World War I, the Livens Drum was produced at Army facilities, such as the Edgewood Arsenal in Maryland and Camp American University in Washington, D.C.

[3] The U.S. Army created five classifications for buried chemical warfare materiel at non-stockpile sites to guide cost and resource requirements for remediating a particular site. Classification data were obtained through site visits, examination of records, interviews, and physical assessment of the sites. Class 1 indicates that the existence of buried chemical warfare materiel has been confirmed by a site assessment or actual recovery. Class 2 indicates that the knowledge of buried chemical warfare materiel is based on documents or interviews. Class 3 indicates that the presence of buried chemical warfare materiel is strongly suspected, based on documents indicating that chemical training, testing, and disposal activities occurred at these sites. Class 4 indicates that buried chemical warfare materiel might be found, based on evidence of past agent manufacturing, storage, or training. Class 5 indicates that a site has been assessed and that no further activity is required or that the site is no longer accessible.

- *Recovered chemical weapons* include items recovered during range-clearing operations, from chemical burial sites, and from research and development testing.[4]
- *Former production facilities* include government facilities that produced chemical weapons and agents prior to the signing of the Chemical Weapons Convention. These facilities produced chemical agent, precursors, and components for chemical weapons or were used for loading and filling munitions.
- *Miscellaneous chemical warfare materiel* includes unfilled munitions, support equipment, and devices designed to be used with chemical weapons. These include complete assembled rounds without chemical fill, with or without bursters and fuzes; simulant-filled munitions; inert munitions; dummy munitions; bursters and fuzes; empty rocket warheads and motors; projectile cases; and other components of metal and plastic parts.

DISPOSAL TECHNOLOGIES, SYSTEMS, AND FACILITIES

The NSCMP is developing a number of technologies, systems, and facilities for identifying, storing, and treating the contents of recovered chemical warfare materiel. The portable isotopic neutron spectroscopy (PINS) device has the capability of identifying the contents of unopened munitions. The presence and relative concentration of a specific chemical element can be determined from characteristic emitted gamma-ray peaks. PINS uses a neutron source, a gamma ray detector, and a multichannel analyzer to identify the chemical elements. A neutron source located near the item being analyzed penetrates the munition's shell and interacts with its contents. The gamma ray detector and multichannel analyzer monitor the energies and intensities of the released gamma rays.

The Raman spectrophotometer analyzes the contents of chemical agent identification sets (CAIS), which consist of chemical agents in glass ampoules, vials, and bottles packed in metal shipping containers or wooden boxes. The Raman spectrophotometer uses a laser to penetrate the glass vials or bottles and identify the contents.

The Mobile Munitions Assessment System (MMAS) is a transportable commercial center that provides on-site information about the contents of unopened recovered munitions and distributes the information to the appropriate authorities and emergency personnel. The MMAS is capable of assessing recovered munitions on site without moving the materiel and also monitoring air at the site. It can determine the contents and stability of either conventional or chemical-filled unopened munitions. Munitions are then analyzed by the PINS. A portable x-ray device may also be used to determine the presence of internal explosive materiel. The MMAS also contains a weather monitoring system. If a leaking munition is present, the weather equipment helps determine the safe evacuation zones away from the site. Cameras are used to monitor all activity around the site. Because the MMAS is powered by a portable gas generator, it can remain on site for

---

[4]Recovered chemical warfare materiel is overpacked and either stored on site or transported and stored at a permitted Department of Defense site following recovery from range-clearing operations and burial. After identifying the type and the quantity of recovered materiel at a site, the Non-Stockpile Chemical Materiel Program (NSCMP) conducts a destination analysis to support the decision to transport or store the materiel. If the decision is to store it on site, the NSCMP prepares an Interim Holding Facility Plan. If the materiel is to be moved for storage and ultimate destruction, the Army prepares a Transportation Plan. The NSCMP considers risk to the public and the environment in deciding on the storage or transportation of the materiel. As required by law, the U.S. Department of Health and Human Services reviews the plans and recommends precautionary measures to protect public health and safety.

months. Redundant computer systems provide data protection in the event of equipment failure, and a backup battery system ensures that no data are lost. The system can be transported by a C-141 cargo aircraft, if necessary, and then driven to a site.

The entire MMAS system is equipped to provide access to sites with varying types of terrain. Once at a site, the full system can be set up in as little as 25 minutes. All communications, photographs, video, x-ray pictures, and computer data can be transmitted immediately via onboard satellite link, cellular phone, and short-wave radio to ensure that the responsible Army officials, state regulators, and local emergency personnel have access to the information and can take necessary actions. Upon completion of the assessment, the MMAS is equipped to decontaminate protective gear and suits, if necessary. Currently, a fully functional MMAS has been fabricated and provided to the Army's Technical Escort Unit for use at Aberdeen Proving Ground, Maryland.

Interim holding facilities (IHFs) provide temporary storage for recovered non-stockpile chemical materiel at sites where permanent storage facilities, such as igloos and bunkers, are not available. An IHF is constructed of fireproof, corrosion-resistant materials, and all electrical fixtures and heating, ventilation, and air conditioning systems are designed to reduce the risk of fire. A secondary containment area below the floor collects any leaking material inside the IHF.

The Rapid Response System (RRS) has the capability of receiving, containing, characterizing, monitoring, and treating (or repackaging) recovered CAIS. The RRS consists of an operations trailer and a utility trailer. Chemical operations, including repackaging and agent neutralization, take place in the glove box, which is housed in the operations trailer. Air circulating through the glove box is vented through charcoal filters to trap agent or other industrial chemicals prior to the discharge of air from the trailer. Air monitoring instruments are also housed inside the trailer. The utility trailer contains electrical generators and other support equipment. Once an agent has been treated, the neutralization wastes are transported to a commercial hazardous waste treatment facility for disposal. Industrial chemicals are also repackaged for transportation to a commercial waste treatment facility.

The Munition Management Device (MMD) has the capability of receiving, containing, accessing, and monitoring buried chemical warfare materiel other than CAIS. Recovered items may be bombs, artillery projectiles, or vials or bottles of agent of various configurations. The MMD consists of two trucks, one for processing and one for control. A munitions treatment vessel, which provides containment of liquid and vapors, tools for accessing the chemicals, and a decontamination solution, is the major component of the system. Other components include a liquid reactor system, a gas reactor/recycling system, a control room, a process laboratory, and a standby generator. The chemical treatment process entails three basic steps: (1) accessing the chemical container; (2) analyzing the chemical; and (3) neutralizing the chemical using a decontamination solution.

D

Legal Context for CAIS Disposal

INTRODUCTION

The National Research Council (NRC) Committee on Review of the Non-Stockpile Chemical Materiel Disposal Program has been requested to review and provide recommendations on the Army's plans for disposing of chemical agent identification sets (CAIS), a relatively small component of the Non-Stockpile Chemical Materiel Disposal Program. These plans are described in the Army's report to Congress, which concludes in part that the "law and its interpretation impose the major obstacles limiting options for the transportation and disposal of CAIS" (U.S. Army, 1998, p. iv). The report also states that the Army has interpreted the legal and regulatory restrictions to mean that "CAIS sets [sic] cannot be processed commercially as hazardous waste due to the current statutory and regulatory framework for handling chemical agent and munitions" (U.S. Army, 1998, p. iii). These broad legal interpretations are a major reason that disposal costs are estimated to be very high.

One of the primary factors that the Army believes contributes to the high costs of CAIS disposal is the legal restriction on the use of federal funds to transport or dispose of lethal chemical warfare agents (50 U.S.C. § 1512). This restriction precludes the transport to or from any military installation or the disposal of chemical warfare agents in the United States until the following actions have been taken:

- U.S. Department of Defense (DoD) has determined that the transportation or testing is in the interests of national security.
- DoD has notified the secretary of health and human services, who has reviewed the potential hazards to public health and safety and recommended precautionary measures.
- DoD has implemented the precautionary measures recommended above. Even if the recommendation prevents the proposed transportation or disposal, the President may determine that overriding considerations of national security require that transportation or disposal proceed.
- DoD has notified: (1) the president of the Senate and the speaker of the House of reesentatives at least 10 days before transportation commences and at least 30 days before testing or disposal commences; and (2) the governor of any state through which chemical warfare agents will be transported.

Similarly, 50 USC § 1512a requires that "chemical munitions" that are "not part of the chemical weapons stockpile" be transported to the nearest chemical munitions stockpile storage facility that has appropriate permits for receiving the item if (1) the secretary of defense considers transportation necessary; and (2) transportation will not pose a significant risk to public health and safety.

In disposing of chemical warfare materiel (CWM), the Army must use "adequate and safe facilities" designed "solely for the destruction of lethal chemical agents and munitions" (50 U.S.C. 1521(c)(1)(B)). However, these facilities may not be used for any purpose "other than the destruction of lethal chemical weapons and munitions" (50 U.S.C. 1521(c)(2)).

Overarching Perspective on Changing the Law

Both the Non-Stockpile Chemical Materiel Program and the Chemical Stockpile Disposal Program must be conducted within a complex legal and regulatory framework that has delicate international implications. From a regulatory point of view, some of the questions are familiar (e.g., the level of acceptable risk), and some are unique (e.g., the required notification of Congress and the affected state[s]).

The regulation of some chemicals varies depending on how they are used. For example, phosgene in a chemical weapon is regulated as a chemical warfare agent. Phosgene in a container at a facility may be regulated as non-stockpile chemical materiel or a toxic chemical ("industrial chemical") subject only to commercial disposal requirements. Interpretation of the legal status of CAIS, which dictates the disposal method, is based more on the judgment of the Army's Office of General Counsel than on explicit language in the statutes.

The safe disposal of non-stockpile chemical agents raises important environmental policy issues, some of which are listed below:

- What is the appropriate process for determining the type and location of disposal?
- How transparent should the decision-making process be?
- What concentration of these chemical agents in ambient air is considered "safe"?
- What are the equities of disposing of chemical agents in one state when they were discovered in a different state?
- What are the appropriate roles of federal, state, and local authorities?

Both the stockpile and non-stockpile chemical materiel disposal programs have also been affected by the military's administrative and organizational framework, which is distinctly different from the commercial waste disposal framework.

CURRENT INTERPRETATION OF CAIS

To understand the dilemma facing the Army requires an understanding of the applicable legal framework, the Army's policy on the disposal of CAIS, and alternative legal interpretations. Several statutory provisions have been enacted in the last several years to implement the stockpile and non-stockpile chemical materiel (disposal) programs:

- The Army currently interprets the statutory scheme to mean that "CAIS cannot be handled commercially as hazardous waste" (U.S. Army, 1998, p. iii) because CAIS are considered CWM (U.S. Army, 1997, p. 1).

APPENDIX D

- The Army interprets 50 U.S.C. § 1512a (which applies to "chemical munitions" that are not part of the chemical weapons stockpile) as precluding the movement of CAIS to commercial treatment, storage, and disposal facilities (U.S. Army, 1997, p. 5).
- CAIS are in included in the definition of recovered CWM in Army Regulation 50-6.
- The Army has concluded that "mustard and Lewisite in any form, including CAIS, will be considered 'lethal chemical agents and munitions' for the purpose of destruction" (U.S. Army, 1998, p. iv).
- The Army interprets the intent of Congress to be that the stockpile chemical agent destruction facilities not be used to destroy CAIS (U.S. Army, 1998, p. 4).

Statutory Definitions

CAIS are difficult to categorize pursuant to the definitions in the Chemical Weapons Convention (CWC) and existing statutes. The difficulty is compounded by the lack of clarity in the congressional language and the fragmentary development of this statutory scheme. Several similar, but different, statutory definitions could apply to CAIS and, indeed, to much of the non-stockpile and stockpile chemical agents.

Similar, but slightly different, terms (lethal chemical warfare agent, lethal chemical agents and munitions, lethal chemical weapons and munitions, chemical agents and munitions, chemical munitions, chemical weapons, toxic chemicals, chemical warfare materiel, and chemical agents) are used in the relevant or related statutes or guidance. Unfortunately, Congress has not specifically defined some of these terms (e.g., lethal chemical warfare agent and chemical munitions) making it difficult to decipher their meaning or Congress' intent.

However, some definitions are included in the statutes or can be deduced from the statutory language. A *chemical agent and munition* means "an agent or munition that, through its chemical properties, produces lethal or other damaging effects on human beings, except such term does not include riot agents, chemical herbicides, smoke and other obscuration materials" (50 U.S.C. § 1521(j)(1)). The term *lethal chemical agent and munition* means "a chemical agent or munition that is designed to cause death, through its chemical properties, to human beings in field concentrations" (50 U.S.C. § 1521(j)(2)). Congress seemed to indicate that the term "any lethal chemical ... agent" was "not intended to apply to use of ... chemical ... materials used for ... test evaluation."[1] Similarly, *chemical warfare materiel* is defined in Army guidance documents as an "item configured as a munition containing a chemical substance that is intended to kill, seriously injure, or incapacitate a person through its physiological effects" (U.S. Army, 1997, p. 1).

In the CWC Implementation Act of 1998, a *chemical weapon* is defined as follows:

- a toxic chemical and its precursors, except where intended for a purpose not prohibited under this chapter as long as the type and quantity is consistent with such a purpose
- a munition or device specifically designed to cause death or other harm through toxic properties of those toxic chemicals specified in subparagraph (a) which would be released by a result of the employment of such munition or device.

---

[1]Conference Report to 50 U.S.C.§ 1512 (1969), cited in Amr et al., 1998, p. 3-3.

- any equipment specifically designed for use directly in connection with the employment of such munition or device[2]

Toxic chemical means any "chemical which through its chemical action on life processes can cause death, temporary incapacitation or permanent harm to humans or animals. The term includes all such chemicals, regardless of their origin or of their method of production, and regardless of whether they are produced in facilities in munitions or elsewhere."[3]

Therefore, it appears that a chemical agent is any substance that produces lethal or other damaging effects on humans. *Lethal chemical agent* appears not to be intended to include chemical agents used for testing. The definition of *munition* or *warfare materiel* seems to be inextricably linked to the intent in designing the device.

There is no debate that (1) the chemicals in CAIS are toxic chemicals; (2) these chemicals can cause temporary incapacitation or permanent harm to humans under some exposure scenarios; (3) CAIS were designed for training purposes, not warfare; (4) CAIS are not munitions; and (5) the volume of, and therefore the risk from, the toxic chemicals in CAIS is less than that of the other materiel in the stockpile and non-stockpile chemical materiel (disposal) programs.

Thus, a possible interpretation is that CAIS are not *lethal chemical warfare agents* or CWM. In this confusing statutory framework and for reasons not apparent in Army policy documents, the Army has strictly construed the statutory scheme and seems to treat CAIS as a *lethal chemical warfare agent* and/or CWM. The implication is that CAIS should be considered a munition or warfare materiel because the contents could be converted into a weapon. The Environmental Protection Agency (EPA) simply cross-references the definition of chemical agents and munitions in 50 U.S.C. section 1521(j)(1)(EPA, 1997, p. 6624). This confusion has impeded the progress of CAIS disposal.

Hazardous Waste Regulations

The Resource Conservation and Recovery Act (RCRA) provides cradle-to-grave regulation of hazardous waste, including the generation, storage, transportation, disposal, and treatment. RCRA includes a corrective action program that requires the cleanup of past or present contamination at the nearly 5,000 or so facilities nationwide that handle, store, transport, or dispose of hazardous waste (in essence, facilities with RCRA permits or interim status). Thus, if CAIS are also interpreted to be hazardous wastes, they must be handled, transported, disposed of, or treated in compliance with EPA's hazardous waste regulations. These requirements, therefore, would dictate the minimum requirements for the storage, disposal, and treatment of CAIS.

EPA has technically reviewed the chemical agents in the stockpile disposal program (the same chemicals that are in the non-stockpile program) and has concluded that these agents, including those in CAIS, are hazardous wastes because they exhibit at least one of

---

[2] Section 2291; Public Law 105-277, Conference Report in H.R. 4328, Making Omnibus Consolidated and Emergency Supplemental Appropriations For Fiscal Year 1999, Congressional Record, H 11277 (daily edition Oct. 19, 1998).

[3] Section 3(10); Public Law 105-277, Conference Report in H.R. 4328, Making Omnibus Consolidated and Emergency Supplemental Appropriations for Fiscal Year 1999, Congressional Record, H11274–11275 (daily edition Oct. 19, 1998).

the characteristics described in EPA's hazardous waste regulations.[4] EPA also observed that chemical agents "are more akin to other types of chemical wastes that RCRA typically regulates than are conventional weapons" (which are also covered by the munitions rule) (EPA, 1997, p. 6638). EPA representatives confirmed to the committee the interpretation that recovered CAIS should be considered hazardous waste. Thus, EPA already considers that the federal hazardous waste regulatory disposal requirements are appropriate for handling chemical agents safely. Therefore, nothing in the federal hazardous waste regulations prohibits the processing of CAIS commercially as hazardous waste.

Requirements for the disposal of hazardous wastes may not be specified in the technical requirements in regulations. In some circumstances, location-specific or chemical-specific permit conditions are developed and made legally binding through the permit process. In other cases, permit writers use guidance and policy documents to add requirements. Unique permit conditions are appropriate if a site-specific risk evaluation indicates that the condition is necessary to protect human health or the environment.

The question of whether unique permit conditions are legally appropriate depends on the law and the record developed during the permit process. Although site-specific requirements are typically resource intensive because they require development on a site-by-site basis, they can be developed and have commonly been included in hazardous waste site permits.

Therefore, although CAIS may be stored, disposed of, or treated at a federally permitted hazardous waste treatment, disposal, or storage facility, additional requirements may also be imposed because of the specific characteristics of this waste.

Permit Modification Requirements

Normally, the type of hazardous waste that can be treated at a commercial facility is limited by the operating permit. Even without an explicit limitation in a permit, a prudent treatment, disposal, or storage facility operator may decide to obtain the approval of the permitting authority prior to accepting a unique type of waste. In some cases, the facility operator may seek objective verification from the waste generator that the waste can legally be disposed of without additional requirements and an ironclad indemnification from the generator.

Thus, EPA or a state may impose a site-specific permit condition (e.g., a limit on the concentration of arsenic in wastes that may be incinerated) if there is a legal and factual basis for such a condition in the administrative record (e.g., a site-specific risk assessment demonstrating that exposures to workers or residents beyond the site boundary would be exposed over a lifetime to unacceptable risks from incineration of wastes containing higher concentrations of arsenic). Therefore, the generator (in this case the Army) will have to supply an objective, scientific record as a basis for allowing treatment of the chemicals in CAIS. Most (but not all) of these chemicals are not "typical" commercial hazardous waste streams. The disposal or storage of CAIS according to the standard operating permit and monitoring requirements may or may not be as "safe" as the disposal of other commercial hazardous wastes. As noted in Chapter 4 of this report, the

---

[4]See EPA, 1997, pp. 6622, 6633, 6638. A substance can be a hazardous waste, among other reasons, because: (1) it is listed as such by EPA in its regulations; or (2) it meets one of the characteristic tests in EPA's regulations (e.g., it meets the criteria for being corrosive or ignitable, or it leaches chemicals at higher levels than specified in the leaching test). See EPA, 1997, p. 6638, and 40 C.F.R. § 261, subpart C. CAIS are not listed as hazardous wastes.

scientific support for allowing the disposal of CAIS at commercial hazardous waste facilities has simply not been developed and is not well documented.

In the unlikely event that neither federal nor state officials requires a permit modification, local and national environmental groups concerned about the disposal of stockpile and non-stockpile agents are likely to petition EPA and/or the state for a permit modification. As a practical matter, therefore, EPA and/or the state regulatory body are likely to require a site-specific permit modification.

MAXIMUM PROTECTION OF HUMAN HEALTH

Introduction

The statutory scheme for the disposal of stockpile chemical agents requires that the Army provide "maximum protection for the environment, the general public, and the personnel who are involved in the destruction of the lethal chemical agents and munitions" (50 U.S.C. § 1521(c)(1)(A)). Although this provision might not apply to CAIS legally, Army guidance documents suggest that it might. There is no clear definition of "maximum protection" or "maximally safe" provided in statute, regulation, or Army guidance. However, there is a wealth of precedent for interpreting similar terms in other statutes.

Overview of Risk Management

Courts give deference to agency decisions that "must be made on the frontiers of science," (i.e., in areas where harm cannot be demonstrated based solely on scientific proof) (50 U.S.C. § 1521(c)(1)(A)). Thus, the U.S. regulatory agencies are generally "not required to support . . . [their] findings with anything approaching scientific certainty" (*Indus. Union Dept. v. API*). Courts in the United States have generally upheld environmental regulations, even though (1) the method of extrapolating from observed biological effects in animals exposed to high levels to predicted adverse health effects in humans exposed to much lower levels, involves assumptions; and (2) there is a lack of scientific certainty regarding the validity of such assumptions (*Indus. Union Dept. v. API*, p. 656; *NRDC v. EPA,* p. 1165). As the court unanimously explained, "there is no particular reason to think that the actual line of the incidence of harm versus degree of exposure is as assumed" (*NRDC v. EPA,* p. 1165).

Most statutes require a balance of risk, costs, and various other factors. However, if statutes explicitly and unequivocally state that no real risk is allowed (e.g., the now repealed Delaney Clause, which forbade carcinogens at any level from being used as food additives), then zero risk is upheld by the courts.[5] In another judicial decision, the court stated that, although "Congress did not dictate that the EPA engage in an exhaustive, full scale, cost-benefit analysis" when making decisions pursuant to the Toxic Substances Control Act, it "did require EPA to consider both sides of the regulatory equation," and it rejected the notion that the EPA should pursue the reduction of risk at any cost (*Corrosion Proof Fittings v. EPA*).

Thus, EPA does not have unlimited discretion, and decision making must include risk management factors, such as costs, depending on the statute. Economic and political factors are explicitly included in the process of selecting the "acceptable" or "safe" risk

---

[5] See for example, *Les v. Reilly*, p. 968.

level (i.e., risk management). It is generally acknowledged that increasing the margin of safety for low-level risks "can reach a point where, by absorbing resources and energy and impeding innovation and growth, it can do both individuals and society more harm than good. . . . The problem is how to know when to stop . . . [and] how to know when prudence and care become over-reaction or paranoia" (Morgan, 1985, pp. 107, 140).

Past Practice

Risk management criteria have evolved over the last several decades. Historically, most chemicals were originally regulated based on noncarcinogenic effects by reviewing animal studies and applying safety and uncertainty factors. Until the late 1950s, few chemicals were regulated as carcinogens. As public perception grew that zero risk could not be attained, a series of statutes were enacted that required the management, rather than elimination, of those risks. Over time, federal regulators developed the risk assessment process (NRC, 1983, pp. 53–55).

Initially, EPA assumed that any concentration of a chemical found to cause cancer in an animal was capable of increasing the lifetime probability of cancer in humans, and the term *safe* was reserved for chemical concentrations that would not result in any probability of adverse health effects (NRC, 1983, pp. 57–58). In 1979, in response to the practical problems of setting regulatory levels, federal agencies (including EPA) formally adopted the risk-assessment process as a tool for setting regulatory levels (Interagency Regulatory Liaison Group, 1979; The Regulatory Council, 1979; EPA, 1979, pp. 58642, 58660.).

Throughout the 1970s, zero exposure was assumed to be the only safe level for carcinogens because, theoretically, even one molecule of a carcinogen could cause cancer, and safe was equated with no risk of adverse health impacts (EPA, 1979; EPA, 1980). Some people even argued that every individual had the "inherent right to protection"[6] and that zero emissions were the only appropriate basis for setting regulatory standards (BNA, 1988). This draconian interpretation would have required the shutdown of most, if not all, facilities in the utility, steel, mining, synthetic chemical, petroleum, and other industries.[7] EPA and the courts concluded that Congress could not have intended to eliminate virtually every major industry in the United States (EPA, 1979, pp. 58652, 58658; *NRDC v. EPA*, p. 1165).

Other regulations issued in the 1970s were based on the extent to which technology could reduce pollution. In the nuclear regulatory field, for example, the lowest achievable levels were used to define radiation safe levels. In other statutes, best available technology was used to define safe levels.

Present Practice

It is now well established in EPA policy and in law, however, that *safe* is not necessarily the same as *risk free*, and mere exposure is not sufficient to support regulation unless there is a significant risk (*Indus. Union Dept. v. API*, p. 642; *NRDC v. EPA*,

---

[6]Natural Resources Defense Council, Supplemental Comments of NRDC on the Environmental Protection Agency National Emissions Standards for Hazardous Air Pollutants: Proposed Standards for Benzene, 1989, p. 4, cited in Marchant and Danzeisen, 1989, pp. 535, 543.

[7]EPA, 1979, p. 58660, as cited in *NRDC v. EPA*, pp. 1146, 1155; Marchant and Danzeisen, 1989, p. 537.

pp. 1164–65). EPA has concluded that a 10^{-4} risk level is to be used at Superfund sites,[8] in national drinking water standards (EPA, 1992, pp. 31797; 31816; 31843),[9] in the Clean Air Act (EPA, 1990b; NRDC v. EPA), in the underground injection control program (EPA, 1988a, pp. 28118; 28123), and in numerous other EPA and other federal regulatory decisions (EPA, 1988b, p. 28486; Wilson and Crouch, 1987; Travis and Hattemer-Frey, 1988; Travis et al., 1987). As a practical matter, technical feasibility, costs, and other factors regularly result in risk levels higher than 10^{-6} being accepted. The "average" level of residual risk considered acceptable by federal agencies in regulatory actions is approximately 10^{-5} (Travis and Hattemer-Frey, 1988, p. 875; Travis et al., 1987, p. 419; EPA, 1990b, pp. 8299–8300).

In all of these regulatory decisions, the agencies and courts eventually arrived at interpretations recognizing that the costs of zero risk are astronomical and, as a practical matter, did not provide a biologically meaningful decrease in the level of the risk. Since the 10^{-4} risk level is safe, there is arguably no meaningful, incremental reduction in human health below the risk level.

Thus, the Army could interpret maximum protection of public health as 10^{-4} risk level. Even though this interpretation is possible, however, nothing compels the interpretation. Obviously, the preferred approach is for Congress to clarify what this term means and then provide adequate funding to implement this level of protection.

LEGAL ISSUES IN CAIS RECOVERY AND DISPOSAL

CAIS Discovery

No specific regulation or guidance document requires a search for CAIS. CAIS are usually found incidentally during normal construction activities on active or former military bases or during investigations initiated at a base as part of an environmental restoration program, a Base Realignment and Closure Act (BRAC) program, or during cleanup of a Superfund site.

On environmental restoration sites, BRAC bases, Superfund sites, and military bases subject to state investigations, efforts to locate CAIS are likely. Otherwise, CAIS may remain buried and may be periodically discovered even after the disposal deadlines in the CWC have passed. In fact, at this time, the Army has no active program to search for CAIS. Once CAIS are found, land use dictates the response. CAIS found in a residential area present different risks than CAIS found on an active military base, a BRAC site, or an industrial or commercial site.

Once found, CAIS are considered CWM and are subject to Army, EPA, and state regulations. Army regulations require the disposal of CWM. CAIS are also considered hazardous waste and must be handled according to federal and state hazardous waste regulations. CAIS are not, however, military munitions as defined by the EPA military munitions rule. EPA, a state, or other regulatory body can impose additional health and safety requirements on permitted facilities receiving CAIS for treatment.

EPA distinguishes between wastes left *in situ* and wastes that have been excavated and moved out of the area of contamination at a Superfund or hazardous waste corrective-action site. No federal, state, or local permits are required for wastes

---

[8] 40 C.F.R. § 300.430(e)(2)(i)(A)(2); EPA, 1990a; *Ohio v. EPA*, pp. 20075–20076.

[9] For example, EPA's drinking water regulations state that 0.2 parts per billion ("ppb") of benzo[a]pyrene in drinking water (the 10^{-4} risk level) "is associated with little to none of . . . [the] risk [from high levels of exposure] and should be considered safe with respect to benzo(a)pyrene." (EPA, 1992, p. 31843).

(including CAIS) stored, disposed of, or treated on a Superfund site, although the substantive requirements of the federal and certain state regulatory requirements must be met. However, the substantive requirements of the federal hazardous waste land disposal restrictions and some other hazardous waste treatment requirements do not apply to waste treated on a Superfund or corrective-action site within what is called a corrective-action management unit (generally, the area of contamination).

On-Site Storage

Once CAIS are removed from the ground and taken outside the corrective-action management unit, they are considered a hazardous waste and CWM. Therefore, CAIS cannot be stored on site for more than 90 days without a permit. If the CAIS are found on a Superfund site, no federal, state, or local permits are required as long as the CAIS remain on the site, although the substantive requirements of the federal and some state environmental laws must be met, unless the remedial action is temporary or one of the other Superfund waivers applies. The technical hazardous waste requirements applicable for long-term storage must be met if CAIS are stored for the long term, unless one of the waiver provisions applies. As a practical matter, the hazardous waste laws require that CAIS be removed from the site in a relatively short period of time.

Treatment in Place

EPA hazardous waste regulations apply to the treatment of CAIS, regardless of whether the treatment is on site, at a regional facility, and regardless of whether a commercial facility or Army facility is used. Presently, CAIS can be treated in place using the Rapid Response System (RRS). However, pursuant to the hazardous waste laws, a hazardous waste cannot be treated at the location where it is found without a hazardous waste permit, unless it is a Superfund site. Thus, the RRS must meet federal and state hazardous waste permitting requirements. EPA or a state can also impose additional health and safety requirements governing the treatment in place of CAIS through the permit process. The Army and other regulatory bodies can also impose additional requirements based on the federal statutes governing CWM.

CAIS Transportation

The movement of CAIS off site is subject to the U.S. Department of Transportation (DOT) rules and Army rules for the transport of CWM. DOT has approved a specific type of container for the transport of CAIS that must be used wherever the CAIS are found and sent by the Army, or possibly by a commercial waste transportation company (DOT, 1998). The RRS also complies with DOT requirements (Cushmac, 1998). According to the Army, current regulations require that the Army Technical Escort Unit transport CAIS to any storage or disposal site.

Off-Site Storage

If CAIS are stored off site, the conditions must comply with EPA hazardous waste storage requirements, whether the facility is commercial or at an Army base. In addition, the Army storage requirements for CWM must be followed. A commercial facility would probably need a permit modification to allow the storage of CAIS. According to the Army, CAIS are currently being stored at Army bases in facilities that meet EPA, state hazardous waste, and Army specifications.

Off-Site Treatment

If CAIS are treated off site, the treatment must be in compliance with the federal and state hazardous waste treatment regulations. A permit modification is likely to be required for commercial facilities.[10] The primary issue for a permit modification is generic (i.e., can CAIS be treated at a permitted hazardous waste treatment facility in a manner that protects human health and the environment).

If a facility does not obtain a permit modification, it faces a legal risk that EPA or the state may take an enforcement action and/or that local citizens may file a citizen action claiming a violation of the permit. Whether public notification prior to accepting CAIS materiel at a commercial facility is required is a complex question. However, failure to notify the public near the commercial facility that the facility is considering accepting CAIS could present significant problems.[11]

EPA, the state, and other regulatory bodies can impose additional health and safety requirements on permitted facilities receiving CAIS for treatment. The Army's policy is not to treat CAIS at stockpile chemical agent treatment facilities, because of 50 U.S.C. 1512a. Any off-site facility or facilities constructed specifically to treat or store CAIS material must meet all hazardous waste regulatory requirements and Army requirements.

Handling Issues

The main difference between the typical treatment of hazardous waste and the treatment of CAIS is in the handling of the CAIS prior to disposal. The most critical issue is the risk of catastrophic failure during unpacking and other handling of the CAIS material. The Army's current procedures seem adequate, but these procedures may not be used at every commercial hazardous waste facility.

REFERENCES

Case Law

Corrosion Proof Fittings v. EPA. Corrosion Proof Fittings versus U.S. Environmental Protection Agency, United States Court of Appeals, Fifth Circuit, No. 89-9596, October 18, 19991. Lexis 2492.
Indus. Union Dept. v. API. Industrial Union Department versus Americal Petroleum Institute, 448 United States Reports.
Les v. Reilly. Les versus Reilly, 968 Federal Reporter, second series, p. 985 (1982).
NRDC v. EPA. Natural Resources Defense Council versus U.S. Environmental Protection Agency, United.States Court of Appeals, D.C. Circuit, 1987. 824 Federal Reporter, second series, pp.1146–ff.

---

[10]Oral communication from EPA headquarters personnel to members of the Committee on Review and Evaluation of the Army Non-Stockpile Chemical Materiel Disposal Program. Actual permitting decisions are typically made by state officials or EPA regional officials.

[11]Recently, DOD attempted to dispose of napalm at an existing commercial hazardous waste treatment facility. Rather than identify the waste as napalm, it was only referred to according to its chemical constituents. The public outcry when the less than full disclosure was uncovered has slowed the disposal process in that case.

Ohio v. EPA. State of Ohio versus U.S. Environmental Protection Agency, U.S. Court of Appeals, D.C. Circuit, 1993. 997 Federal Reporter, Second Series, pp. 1520,1532.

Publications

Amr, A., A. Goldfarb, S. Haus, L. Hourcle, M. Simmons, A. Talib, D. Tripler, R. Wassmann, and A. Wusterbath. 1998. Preliminary Assessment of the Commercial Viability for CAIS Treatment and Disposal. Mitretek Technical Report MTR-1998-5. June 1998. McLean, Va.: Mitretek Systems.

BNA (Bureau of National Affairs). 1988. Benzene rules to heed vinyl chloride decision, though controls may be same, EPA analyst says. BNA Environment Report 18: 2011–2012.

Cushmac. G. 1998. Presentation by George Cushmac, U.S. Department of Transportation, to the Committee on Review and Evaluation of the Army Non-Stockpile Chemical Materiel Disposal Program, October 29, 1998, National Research Council, Washington, D.C.

DOT (U.S. Department of Transportation). 1998. Approval CA-9804018 (April 14, 1998) issued to the Department of the Army, approving the shipment of "recovered chemical warfare materiel" in the packaging described. Washington, D.C.: U.S. Department of Transportation.

EPA (U.S. Environmental Protection Agency). 1979. National emission standards for hazardous air pollutants; policy and procedures for identifying, assessing and regulating airborne substances posing a risk of cancer. Federal Register 44: 58642, 58660.

EPA. 1980. Water quality criteria documents. Federal Register 45: 79318–79323 Interagency Regulatory Liaison Group. 1979. Scientific Basis for Identification of Potential Carcinogen and Estimation of Risks.

EPA. 1988a. Underground injection control program: Hazardous waste disposal injection restrictions; amendments to technical requirements for Class 1 hazardous waste injection wells; and additional monitoring requirements applicable to all Class 1 wells. Federal Register 53: 28118–28123.

EPA. 1988b. National emission standards for hazardous air pollutants; benzene emissions from maleic anhydride plants, ethylbenzene/styrene plants, benzene storage vessels, benzene equipment leaks, and coke by-product recovery plants. Proposed rule. Federal Register 53: 28496.

EPA. 1990a. National oil pollution and hazardous substances contingency plan. Federal Register 56: 8666–ff.

EPA. 1990b. National emission standards for hazardous air pollutants; benzene emissions from chemical manufacturing process vents, industrial solvent use, benzene waste operation, benzene transfer operations, and gasoline marketing system. Final rule. Federal Register 55: 8292, 8299–8300.

EPA. 1992. Drinking water; national primary drinking water regulations: synthetic organic chemicals and inorganic chemicals; national primary drinking water regulations implementation. Final rule. Federal Register 57: 31776–ff.

EPA. 1997. Military munitions rule: hazardous waste identification and management; explosives emergencies; manifest exemption for transport of hazardous waste on rights-of-way on contiguous properties; Final rule. Federal Register 62(Feb. 12): 6621–6657.

Interagency Regulatory Liaison Group.1979. Scientific Basis for Identification of Potential Carcinogen and Estimation of Risk. Washington, D.C.: U.S. Government Printing Office.

Marchant, G., and D. Danzeisen. 1989. "Acceptable" risk for hazardous air pollutants. Harvard Environmental Law Review 13: 535–558.

Morgan, G.1985. Risk assessment and risk management decision making for chemical exposure. Chapter 5 in Environmental Exposure from Chemicals, Vol. 2. Boca Raton, FL: CRC Press.

NRC (National Research Council). 1983. Risk Assessment in the Federal Government: Managing the Process. Committee on the Institutional Means for Assessment of Risks to Public Health. Washington, D.C.: National Academy Press.

Regulatory Council.1979. Regulation of Chemical Carcinogens. Washington, D.C.: U.S. Government Printing Office.

Travis, C., S. Richter, E. Crouch, R. Wilson, and E. Klema. 1987. Cancer risk management: a review of 132 federal regulatory decisions. Environmental Science and Technology 21(5): 415–420.

Travis, C. and H. Hattemer-Frey. 1988. Determining an acceptable level of risk. Environmental Science and Technology 22(8): 873–876.

U.S. Army. 1997. Interim Guidance for Biological Warfare Materiel (BWM) and Chemical Warfare Materiel (CWM) Response Activities. September 5. Washington, D.C.: Department of the Army.

U.S. Army. 1998. Report to Congress on Alternative Approaches for the Treatment and Disposal of Chemical Agent Identification Sets (CAIS). Prepared by the Project Manager for Non-Stockpile Chemical Materiel. June 1998. Aberdeen Proving Ground, Md.: U.S. Army Program Manager for Chemical Demilitarization.

Wilson, R., and E. Crouch. 1987. Risk assessment and comparisons: an introduction. Science 236(April 17): 267–270.